Handbook

Table of Contents

Cells and DNA

Table of Contents

What is a cell?

Cells are the basic building blocks of all living things. The human body is composed of trillions of cells. They provide structure for the body, take in nutrients from food, convert those nutrients into energy, and carry out specialized functions. Cells also contain the body's hereditary material and can make copies of themselves.

Cells have many parts, each with a different function. Some of these parts, called organelles, are specialized structures that perform certain tasks within the cell. Human cells contain the following major parts, listed in alphabetical order:

Cytoplasm (illustration on page 6)

> Within cells, the cytoplasm is made up of a jelly-like fluid (called the cytosol) and other structures that surround the nucleus.

Cytoskeleton

> The cytoskeleton is a network of long fibers that make up the cell's structural framework. The cytoskeleton has several critical functions, including determining cell shape, participating in cell division, and allowing cells to move. It also provides a track-like system that directs the movement of organelles and other substances within cells.

Endoplasmic reticulum (ER) (illustration on page 6)

> This organelle helps process molecules created by the cell. The endoplasmic reticulum also transports these molecules to their specific destinations either inside or outside the cell.

Golgi apparatus (illustration on page 7)

> The Golgi apparatus packages molecules processed by the endoplasmic reticulum to be transported out of the cell.

Lysosomes and peroxisomes (illustration on page 7)

> These organelles are the recycling center of the cell. They digest foreign bacteria that invade the cell, rid the cell of toxic substances, and recycle worn-out cell components.

Mitochondria (illustration on page 7)

> Mitochondria are complex organelles that convert energy from food into a form that the cell can use. They have their own genetic material, separate from the DNA in the nucleus, and can make copies of themselves.

Nucleus (illustration on page 8)

>The nucleus serves as the cell's command center, sending directions to the cell to grow, mature, divide, or die. It also houses DNA (deoxyribonucleic acid), the cell's hereditary material. The nucleus is surrounded by a membrane called the nuclear envelope, which protects the DNA and separates the nucleus from the rest of the cell.

Plasma membrane (illustration on page 8)

>The plasma membrane is the outer lining of the cell. It separates the cell from its environment and allows materials to enter and leave the cell.

Ribosomes (illustration on page 8)

>Ribosomes are organelles that process the cell's genetic instructions to create proteins. These organelles can float freely in the cytoplasm or be connected to the endoplasmic reticulum (see above).

For more information about cells:

The Genetic Science Learning Center at the University of Utah offers an interactive introduction to cells (http://learn.genetics.utah.edu/content/cells/insideacell/) and their many functions.

Nature Education's Scitable explains what cells are made of and how they originated in their fact sheet What is a Cell? (http://www.nature.com/scitable/topicpage/what-is-a-cell-14023083)

Arizona State University's "Ask a Biologist" provides a description and illustration of each of the cell's organelles (http://askabiologist.asu.edu/content/cell-parts).

Queen Mary University of London allows you to explore a 3-D cell and its parts (http://www.centreofthecell.org/interactives/exploreacell/index.php).

Additional information about the cytoskeleton, including an illustration, is available from the Cytoplasm Tutorial (http://www.biology.arizona.edu/Cell_bio/tutorials/cytoskeleton/page1.html). This resource is part of The Biology Project at the University of Arizona.

Illustrations

Cytoplasm

The cytoplasm surrounds the cell's nucleus and organelles.

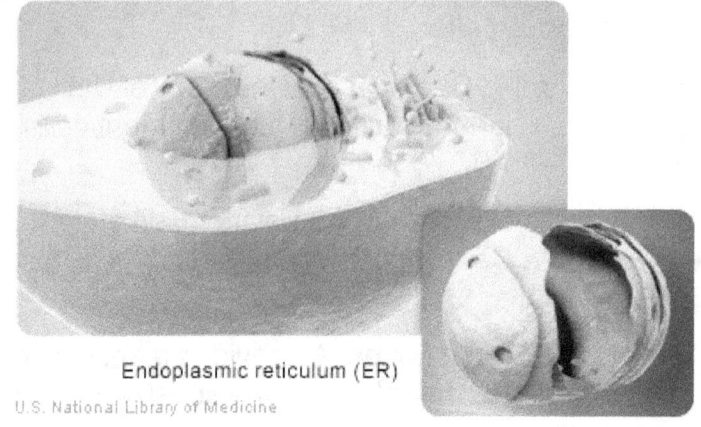

Endoplasmic reticulum (ER)

The endoplasmic reticulum is involved in molecule processing and transport.

Golgi apparatus

U.S. National Library of Medicine

The Golgi apparatus is involved in packaging molecules for export from the cell.

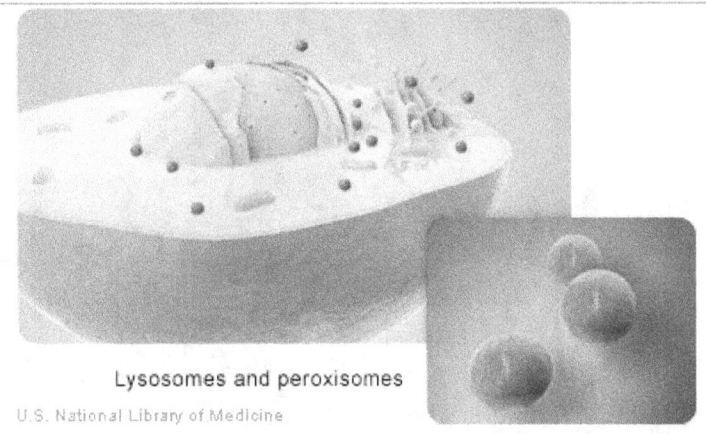

Lysosomes and peroxisomes

U.S. National Library of Medicine

Lysosomes and peroxisomes destroy toxic substances and recycle worn-out cell parts.

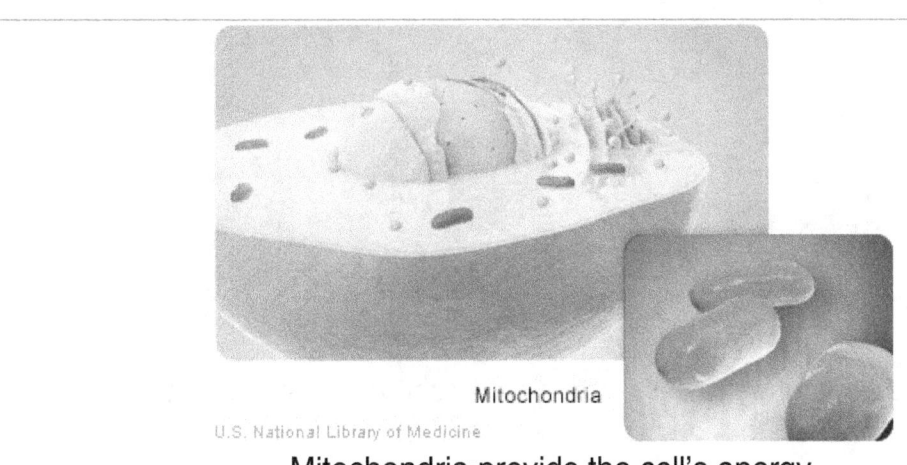

Mitochondria

U.S. National Library of Medicine

Mitochondria provide the cell's energy.

The nucleus contains most of the cell's genetic material.

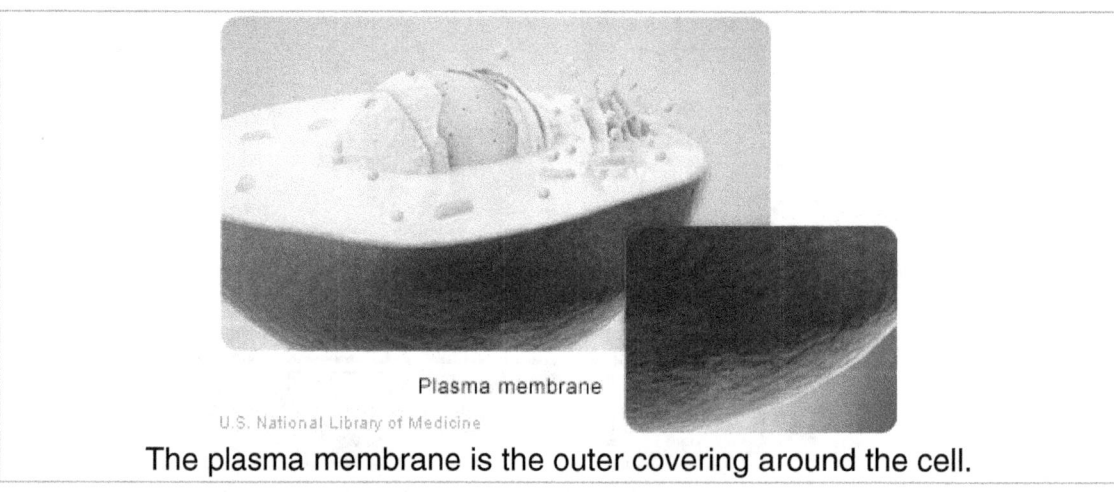

The plasma membrane is the outer covering around the cell.

Ribosomes use the cell's genetic instructions to make proteins.

What is DNA?

DNA, or deoxyribonucleic acid, is the hereditary material in humans and almost all other organisms. Nearly every cell in a person's body has the same DNA. Most DNA is located in the cell nucleus (where it is called nuclear DNA), but a small amount of DNA can also be found in the mitochondria (where it is called mitochondrial DNA or mtDNA).

The information in DNA is stored as a code made up of four chemical bases: adenine (A), guanine (G), cytosine (C), and thymine (T). Human DNA consists of about 3 billion bases, and more than 99 percent of those bases are the same in all people. The order, or sequence, of these bases determines the information available for building and maintaining an organism, similar to the way in which letters of the alphabet appear in a certain order to form words and sentences.

DNA bases pair up with each other, A with T and C with G, to form units called base pairs. Each base is also attached to a sugar molecule and a phosphate molecule. Together, a base, sugar, and phosphate are called a nucleotide. Nucleotides are arranged in two long strands that form a spiral called a double helix. The structure of the double helix is somewhat like a ladder, with the base pairs forming the ladder's rungs and the sugar and phosphate molecules forming the vertical sidepieces of the ladder.

An important property of DNA is that it can replicate, or make copies of itself. Each strand of DNA in the double helix can serve as a pattern for duplicating the sequence of bases. This is critical when cells divide because each new cell needs to have an exact copy of the DNA present in the old cell.

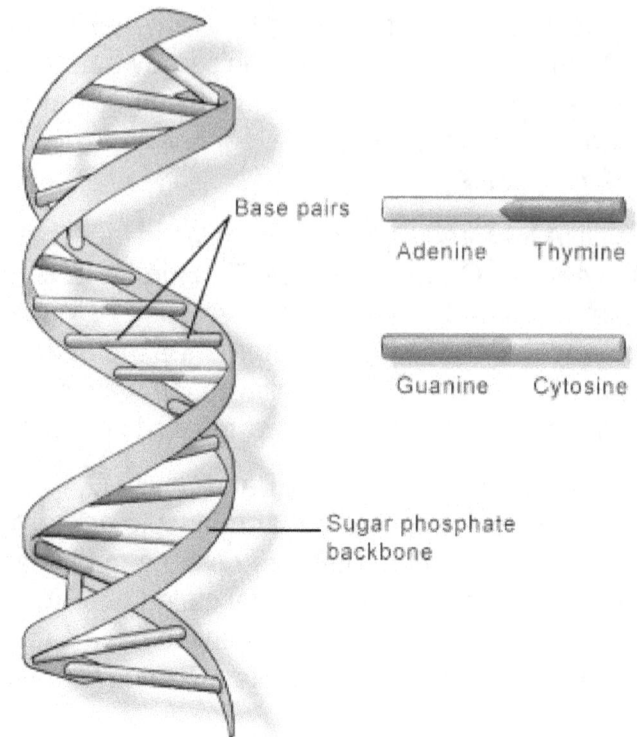

Base pairs

Adenine   Thymine

Guanine   Cytosine

Sugar phosphate
backbone

U.S. National Library of Medicine

DNA is a double helix formed by base pairs attached to a sugar-phosphate
backbone.

For more information about DNA:

The National Human Genome Research Institute fact sheet Deoxyribonucleic Acid
(DNA) (http://www.genome.gov/25520880) provides an introduction to this molecule.

Information about the genetic code (http://geneed.nlm.nih.gov/topic_subtopic.php?
tid=15&sid=19) and the structure of the DNA double helix (http://geneed.nlm.nih.gov/
topic_subtopic.php?tid=15&sid=16) is available from GeneEd.

The New Genetics, a publication of the National Institute of General Medical
Sciences, discusses the structure of DNA and how it was discovered
(http://publications.nigms.nih.gov/thenewgenetics/chapter1.html#c1).

Nature Education's Scitable offers a thorough description of DNA
(http://www.nature.com/scitable/topicpage/DNA-Is-a-Structure-that-Encodes-
Information-6493050), including its components and organization. It also includes
a short animated video.

A basic explanation and illustration of DNA (http://www.nature.com/scitable/topicpage/DNA-Is-a-Structure-that-Encodes-Information-6493050) can be found on Arizona State University's "Ask a Biologist" website.

The Virtual Genetics Education Centre, created by the University of Leicester, offers additional information on DNA, genes, and chromosomes (http://www2.le.ac.uk/departments/genetics/vgec/schoolscolleges/topics/dna-genes-chromosomes).

What is mitochondrial DNA?

Although most DNA is packaged in chromosomes within the nucleus, mitochondria also have a small amount of their own DNA. This genetic material is known as mitochondrial DNA or mtDNA.

Mitochondria (illustration on page 7) are structures within cells that convert the energy from food into a form that cells can use. Each cell contains hundreds to thousands of mitochondria, which are located in the fluid that surrounds the nucleus (the cytoplasm).

Mitochondria produce energy through a process called oxidative phosphorylation. This process uses oxygen and simple sugars to create adenosine triphosphate (ATP), the cell's main energy source. A set of enzyme complexes, designated as complexes I-V, carry out oxidative phosphorylation within mitochondria.

In addition to energy production, mitochondria play a role in several other cellular activities. For example, mitochondria help regulate the self-destruction of cells (apoptosis). They are also necessary for the production of substances such as cholesterol and heme (a component of hemoglobin, the molecule that carries oxygen in the blood).

Mitochondrial DNA contains 37 genes, all of which are essential for normal mitochondrial function. Thirteen of these genes provide instructions for making enzymes involved in oxidative phosphorylation. The remaining genes provide instructions for making molecules called transfer RNAs (tRNAs) and ribosomal RNAs (rRNAs), which are chemical cousins of DNA. These types of RNA help assemble protein building blocks (amino acids) into functioning proteins.

For more information about mitochondria and mitochondrial DNA:

Molecular Expressions, a web site from the Florida State University Research Foundation, offers an illustrated introduction to mitochondria and mitochondrial DNA (http://micro.magnet.fsu.edu/cells/mitochondria/mitochondria.html).

An overview of mitochondrial DNA (http://neuromuscular.wustl.edu/mitosyn.html# general) is available from the Neuromuscular Disease Center at Washington University.

Nature Education's Scitable provides a comprehensive explanation of mitochondrial DNA and the conditions that can be associated with mitochondrial DNA mutations (http://www.nature.com/scitable/topicpage/mtdna-and-mitochondrial-diseases-903).

What is a gene?

A gene is the basic physical and functional unit of heredity. Genes, which are made up of DNA, act as instructions to make molecules called proteins. In humans, genes vary in size from a few hundred DNA bases to more than 2 million bases. The Human Genome Project has estimated that humans have between 20,000 and 25,000 genes.

Every person has two copies of each gene, one inherited from each parent. Most genes are the same in all people, but a small number of genes (less than 1 percent of the total) are slightly different between people. Alleles are forms of the same gene with small differences in their sequence of DNA bases. These small differences contribute to each person's unique physical features.

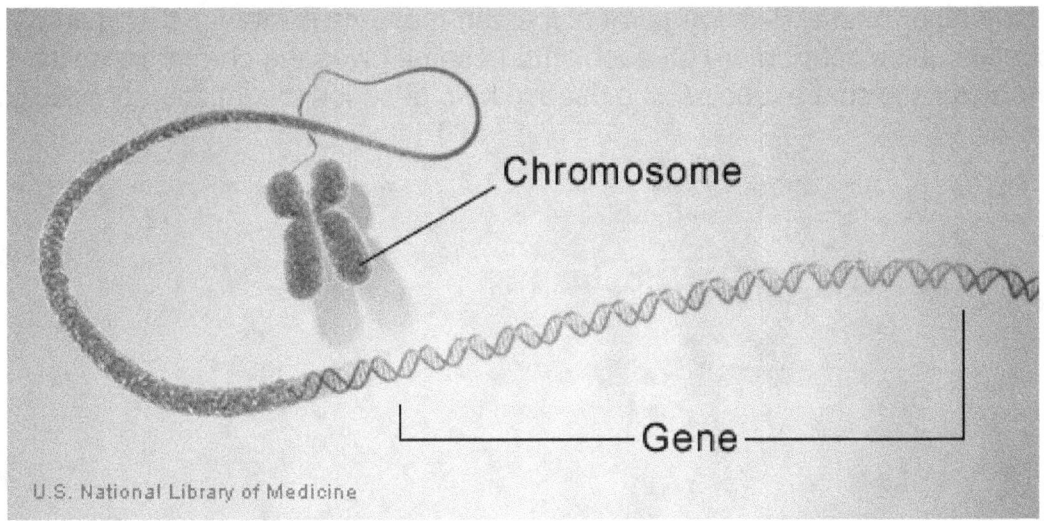

U.S. National Library of Medicine

Genes are made up of DNA. Each chromosome contains many genes.

For more information about genes:

Genetics Home Reference provides consumer-friendly gene summaries (http://ghr.nlm.nih.gov/BrowseGenes) that include an explanation of each gene's normal function and how mutations in the gene cause particular genetic conditions.

The Centre for Genetics Education offers a fact sheet that introduces genes and chromosomes (http://www.genetics.edu.au/Publications-and-Resources/Genetics-Fact-Sheets/FactSheet1GenesandChromosomesTheGenome.pdf).

The Tech Museum of Innovation at Stanford University describes genes and how they were discovered (http://genetics.thetech.org/about-genetics/what-gene).

The Virtual Genetics Education Centre, created by the University of Leicester, offers additional information on DNA, genes, and chromosomes (http://www2.le.ac.uk/departments/genetics/vgec/schoolscolleges/topics/dna-genes-chromosomes).

What is a chromosome?

In the nucleus of each cell, the DNA molecule is packaged into thread-like structures called chromosomes. Each chromosome is made up of DNA tightly coiled many times around proteins called histones that support its structure.

Chromosomes are not visible in the cell's nucleus—not even under a microscope—when the cell is not dividing. However, the DNA that makes up chromosomes becomes more tightly packed during cell division and is then visible under a microscope. Most of what researchers know about chromosomes was learned by observing chromosomes during cell division.

Each chromosome has a constriction point called the centromere, which divides the chromosome into two sections, or "arms." The short arm of the chromosome is labeled the "p arm." The long arm of the chromosome is labeled the "q arm." The location of the centromere on each chromosome gives the chromosome its characteristic shape, and can be used to help describe the location of specific genes.

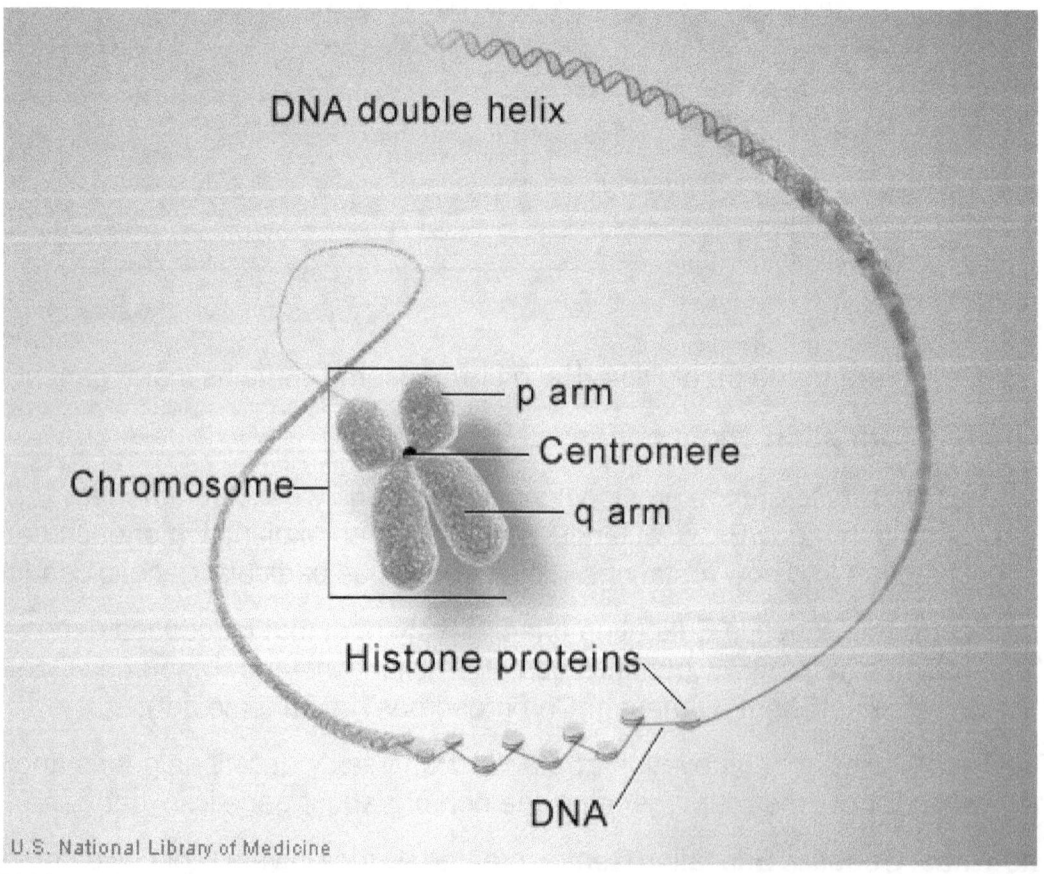

DNA double helix

p arm

Centromere

Chromosome

q arm

Histone proteins

DNA

U.S. National Library of Medicine

DNA and histone proteins are packaged into structures called chromosomes.

For more information about chromosomes:

Genetics Home Reference provides information about each human chromosome (http://ghr.nlm.nih.gov/chromosomes) written in lay language.

The Centre for Genetics Education offers a fact sheet that introduces genes and chromosomes (http://www.genetics.edu.au/Publications-and-Resources/Genetics-Fact-Sheets/FactSheet1GenesandChromosomesTheGenome.pdf).

GeneEd also provides information about the basics of chromosomes (http://geneed.nlm.nih.gov/topic_subtopic.php?tid=15&sid=17).

The University of Utah's Genetic Science Learning Center offers a description of chromosomes (http://www.virtual.unal.edu.co/cursos/ingenieria/2001832/lecturas/chromosome.swf), including their size and how they are packaged in the cell.

How many chromosomes do people have?

In humans, each cell normally contains 23 pairs of chromosomes, for a total of 46. Twenty-two of these pairs, called autosomes, look the same in both males and females. The 23rd pair, the sex chromosomes, differ between males and females. Females have two copies of the X chromosome, while males have one X and one Y chromosome.

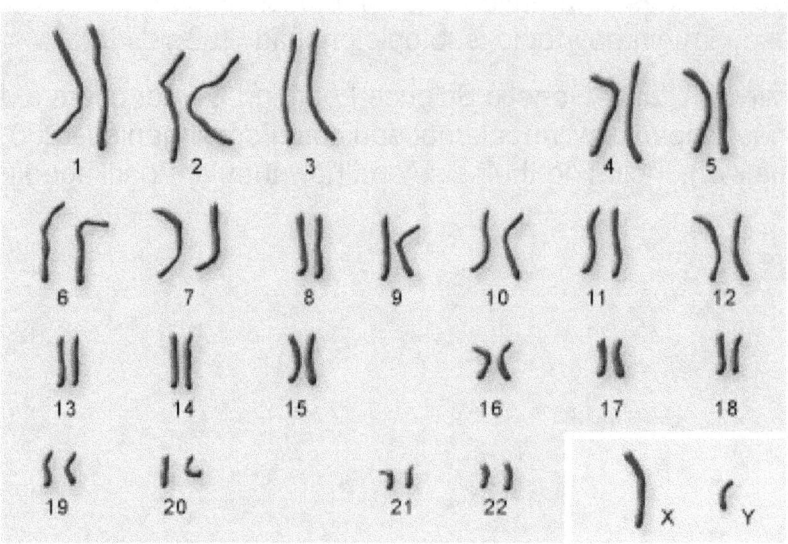

autosomes   sex chromosomes

U.S. National Library of Medicine

The 22 autosomes are numbered by size. The other two chromosomes, X and Y, are the sex chromosomes. This picture of the human chromosomes lined up in pairs is called a karyotype.

For more information about the 23 pairs of human chromosomes:

Genetics Home Reference provides information about each human chromosome (http://ghr.nlm.nih.gov/chromosomes) written in lay language.

The University of Utah's Genetic Science Learning Center discusses how karyotypes can be used in diagnosing genetic disorders (http://learn.genetics.utah.edu/content/chromosomes/diagnose/).

Arizona State University's "Ask a Biologist" discusses the inheritance of human chromosomes. (http://askabiologist.asu.edu/chromosomes-and-genes)

Chapter 2

How Genes Work

Table of Contents

What are proteins and what do they do?

Proteins are large, complex molecules that play many critical roles in the body. They do most of the work in cells and are required for the structure, function, and regulation of the body's tissues and organs.

Proteins are made up of hundreds or thousands of smaller units called amino acids, which are attached to one another in long chains. There are 20 different types of amino acids that can be combined to make a protein. The sequence of amino acids determines each protein's unique 3-dimensional structure and its specific function.

Proteins can be described according to their large range of functions in the body, listed in alphabetical order:

Examples of protein functions

Function	Description	Example
Antibody	Antibodies bind to specific foreign particles, such as viruses and bacteria, to help protect the body.	Immunoglobulin G (IgG) (illustration on page 19)
Enzyme	Enzymes carry out almost all of the thousands of chemical reactions that take place in cells. They also assist with the formation of new molecules by reading the genetic information stored in DNA.	Phenylalanine hydroxylase (illustration on page 20)
Messenger	Messenger proteins, such as some types of hormones, transmit signals to coordinate biological processes between different cells, tissues, and organs.	Growth hormone (illustration on page 21)
Structural component	These proteins provide structure and support for cells. On a larger scale, they also allow the body to move.	Actin (illustration on page 22)
Transport/storage	These proteins bind and carry atoms and small molecules within cells and throughout the body.	Ferritin (illustration on page 23)

For more information about proteins and their functions:

KidsHealth from Nemours offers a basic overview of proteins (http://kidshealth.org/kid/stay_healthy/body/protein.html) and what they do.

Nature Education's Scitable offers information on the function of proteins. (http://www.nature.com/scitable/topicpage/protein-function-14123348)

Arizona State University's "Ask a Biologist" discusses the different kinds of proteins (http://askabiologist.asu.edu/venom/what-are-proteins) and what they do.

Illustrations

Immunoglobulin G (IgG)

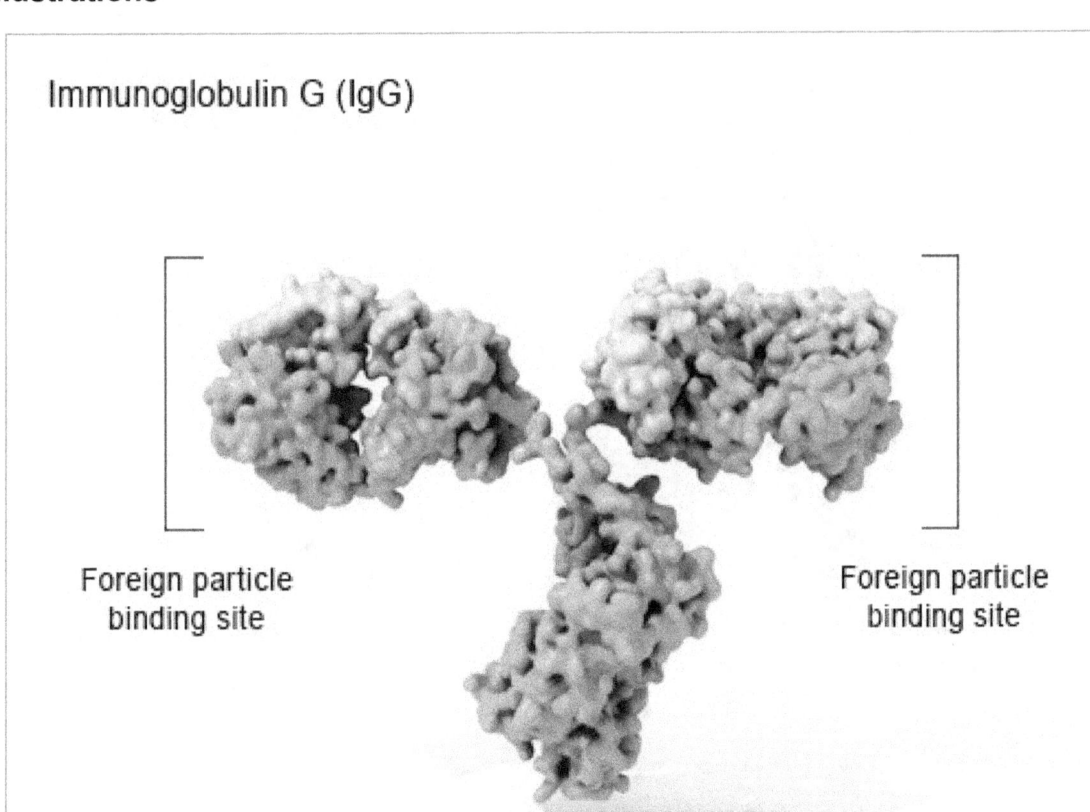

Foreign particle binding site

Foreign particle binding site

Immunoglobulin G is a type of antibody that circulates in the blood and recognizes foreign particles that might be harmful.

Phenylalanine hydroxylase

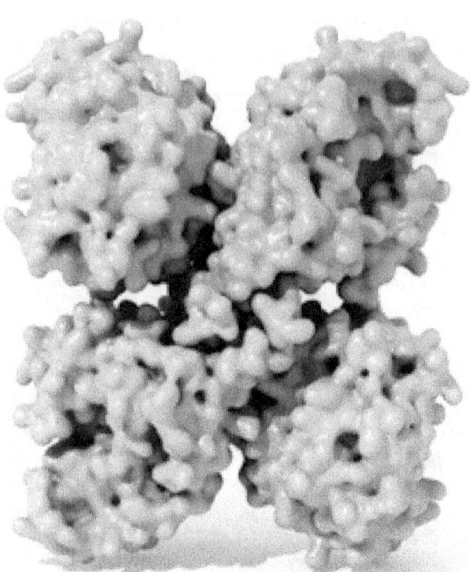

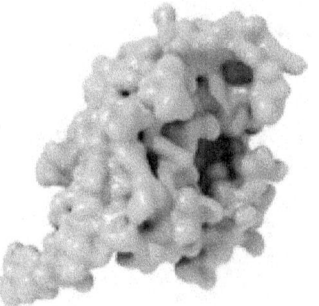

Single phenylalanine
hydroxylase subunit

Phenylalanine hydroxylase
protein consisting of 4 subunits

The functional phenylalanine hydroxylase enzyme is made up of four identical
subunits. The enzyme converts the amino acid phenylalanine to another amino
acid, tyrosine.

Growth hormone

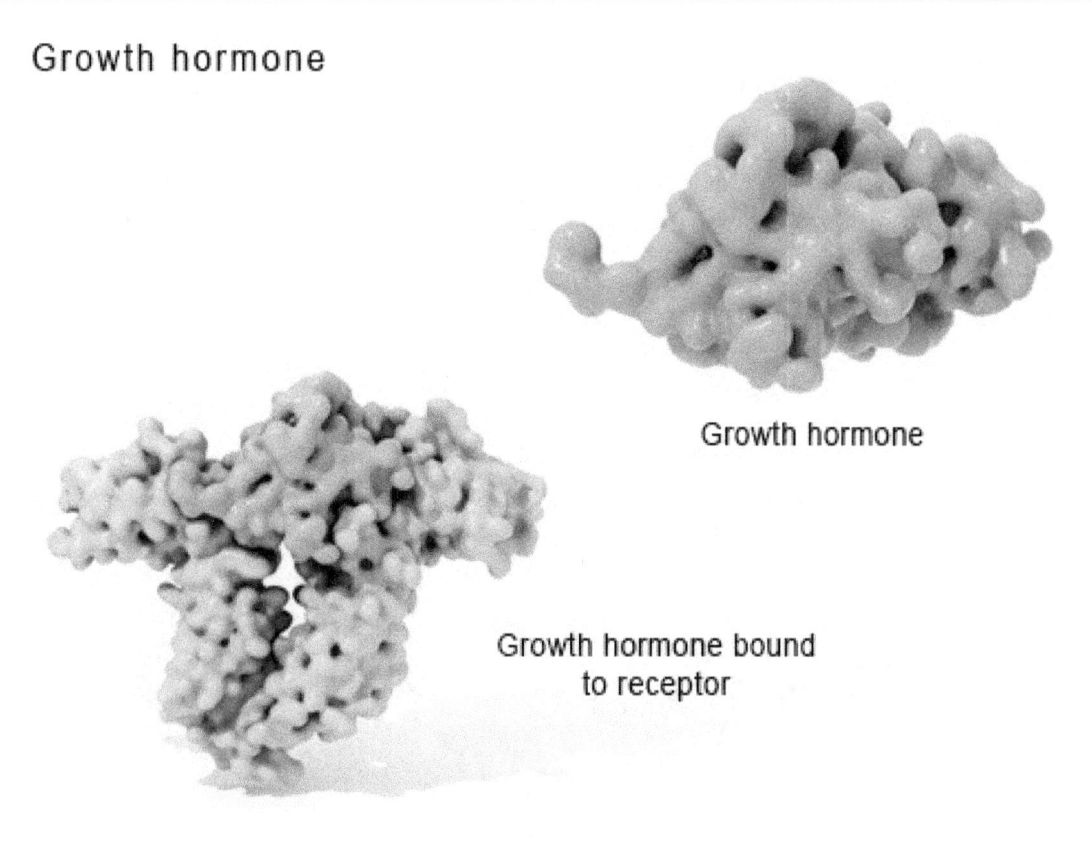

Growth hormone

Growth hormone bound
to receptor

Growth hormone is a messenger protein made by the pituitary gland. It regulates cell growth by binding to a protein called a growth hormone receptor.

Actin

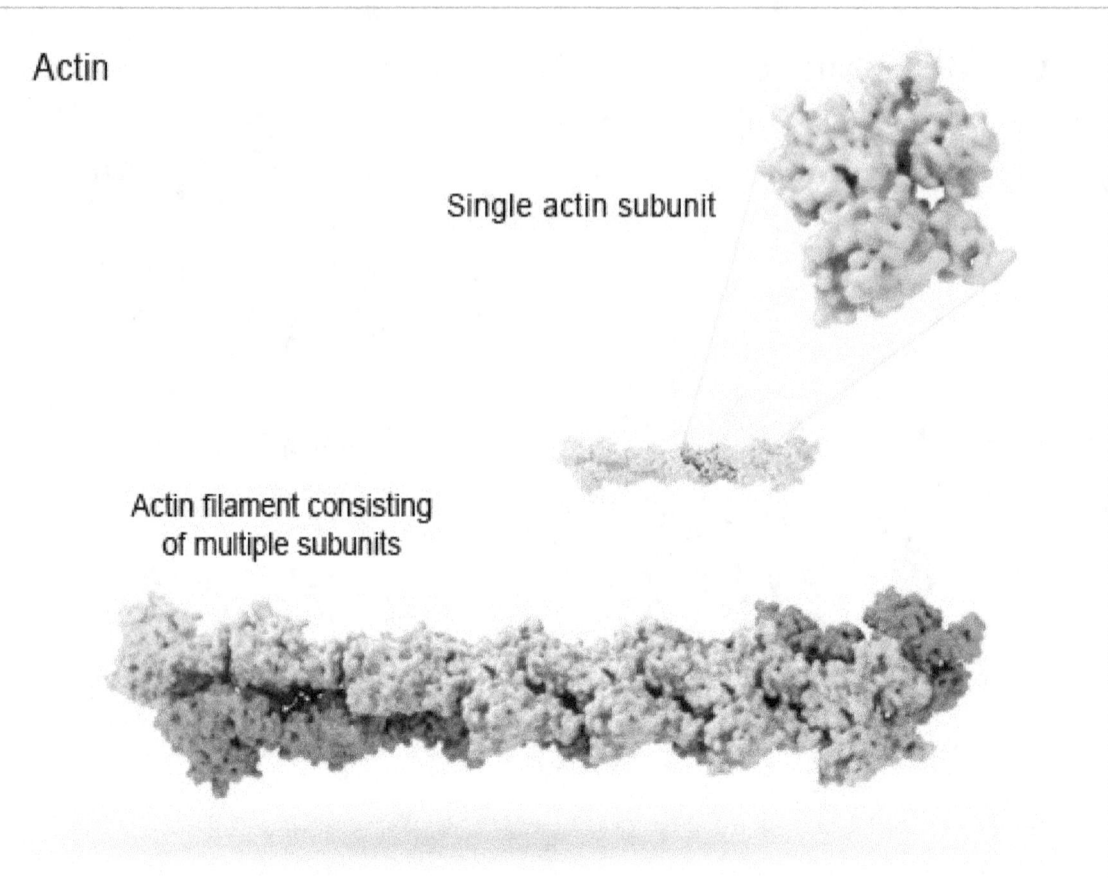

Single actin subunit

Actin filament consisting
of multiple subunits

Actin filaments, which are structural proteins made up of multiple subunits,
help muscles contract and cells maintain their shape.

Ferritin

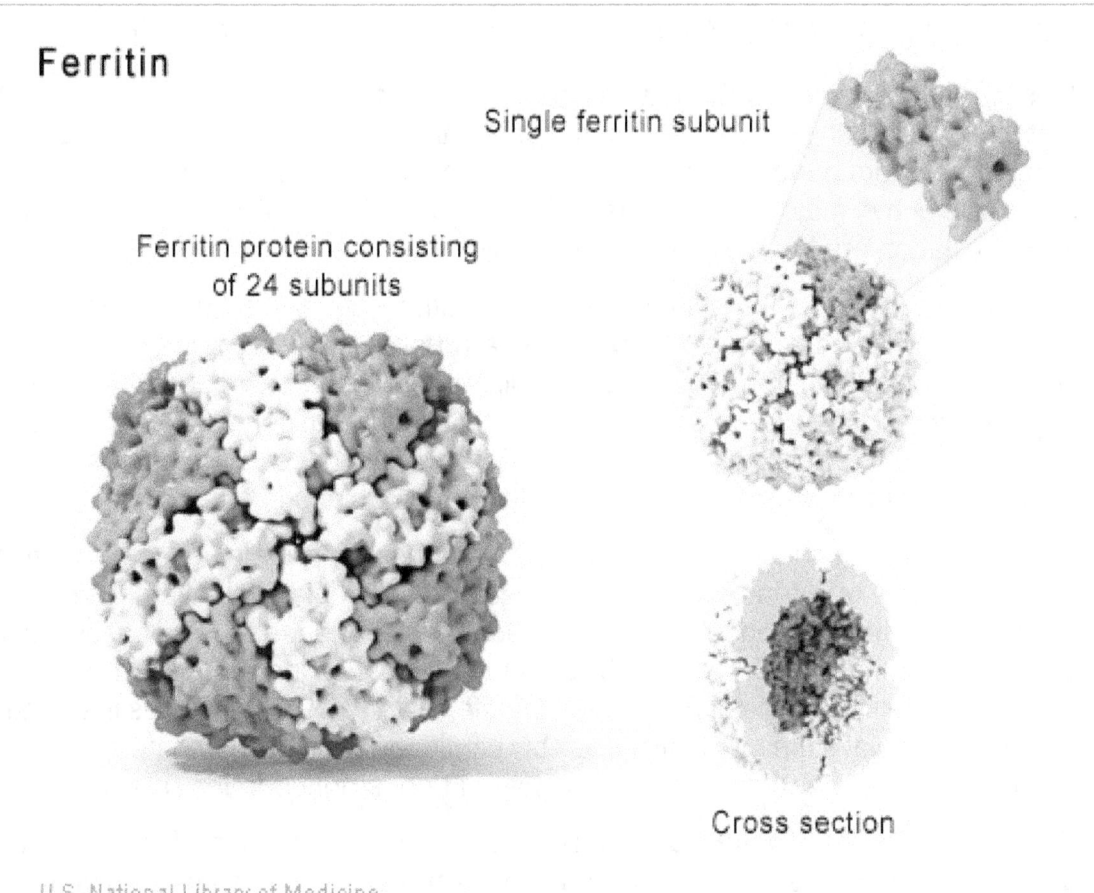

Single ferritin subunit

Ferritin protein consisting
of 24 subunits

Cross section

Ferritin, a protein made up of 24 identical subunits, is involved in iron storage.

How do genes direct the production of proteins?

Most genes contain the information needed to make functional molecules called proteins. (A few genes produce other molecules that help the cell assemble proteins.) The journey from gene to protein is complex and tightly controlled within each cell. It consists of two major steps: transcription and translation. Together, transcription and translation are known as gene expression.

During the process of transcription, the information stored in a gene's DNA is transferred to a similar molecule called RNA (ribonucleic acid) in the cell nucleus. Both RNA and DNA are made up of a chain of nucleotide bases, but they have slightly different chemical properties. The type of RNA that contains the information for making a protein is called messenger RNA (mRNA) because it carries the information, or message, from the DNA out of the nucleus into the cytoplasm.

Translation, the second step in getting from a gene to a protein, takes place in the cytoplasm. The mRNA interacts with a specialized complex called a ribosome, which "reads" the sequence of mRNA bases. Each sequence of three bases, called a codon, usually codes for one particular amino acid. (Amino acids are the building blocks of proteins.) A type of RNA called transfer RNA (tRNA) assembles the protein, one amino acid at a time. Protein assembly continues until the ribosome encounters a "stop" codon (a sequence of three bases that does not code for an amino acid).

The flow of information from DNA to RNA to proteins is one of the fundamental principles of molecular biology. It is so important that it is sometimes called the "central dogma."

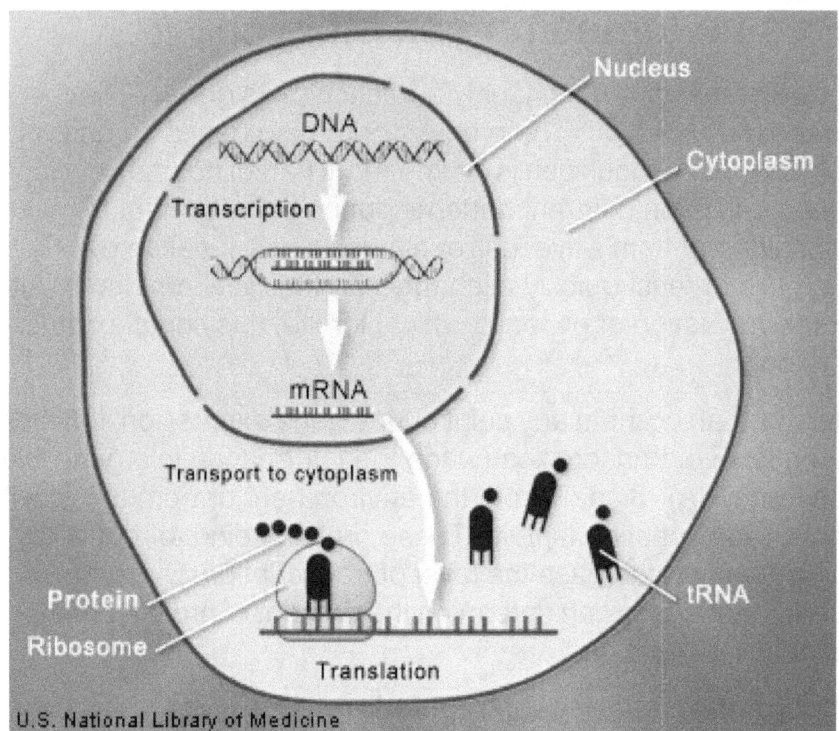

Through the processes of transcription and translation, information from genes is used to make proteins.

For more information about making proteins:

The Genetic Science Learning Center at the University of Utah offers an interactive introduction to transcription and translation (http://learn.genetics.utah.edu/content/molecules/dna/).

Information about RNA (http://geneed.nlm.nih.gov/topic_subtopic.php?tid=15&sid=18), transcription (http://geneed.nlm.nih.gov/topic_subtopic.php?tid=15&sid=22), and translation (http://geneed.nlm.nih.gov/topic_subtopic.php?tid=15&sid=23) is available from GeneEd.

North Dakota State University's Virtual Cell Animation Collection offers videos that illustrate the processes of transcription (http://vcell.ndsu.nodak.edu/animations/transcription/movie-flash.htm) and translation (http://vcell.ndsu.nodak.edu/animations/translation/movie-flash.htm).

The New Genetics, a publication of the National Institute of General Medical Sciences, includes discussions of transcription (http://publications.nigms.nih.gov/thenewgenetics/chapter1.html#c4) and translation (http://publications.nigms.nih.gov/thenewgenetics/chapter1.html#c7).

Can genes be turned on and off in cells?

Each cell expresses, or turns on, only a fraction of its genes. The rest of the genes are repressed, or turned off. The process of turning genes on and off is known as gene regulation. Gene regulation is an important part of normal development. Genes are turned on and off in different patterns during development to make a brain cell look and act different from a liver cell or a muscle cell, for example. Gene regulation also allows cells to react quickly to changes in their environments. Although we know that the regulation of genes is critical for life, this complex process is not yet fully understood.

Gene regulation can occur at any point during gene expression, but most commonly occurs at the level of transcription (when the information in a gene's DNA is transferred to mRNA). Signals from the environment or from other cells activate proteins called transcription factors. These proteins bind to regulatory regions of a gene and increase or decrease the level of transcription. By controlling the level of transcription, this process can determine the amount of protein product that is made by a gene at any given time.

For more information about gene regulation:

The Genetic Science Learning Center at the University of Utah offers an explanation of gene expression as it relates to disease risk (http://learn.genetics.utah.edu/content/science/expression/).

Education Portal provides information about gene regulation in Regulation of Gene Expression: Transcriptional Repression and Induction (http://education-portal.com/academy/lesson/regulation-of-gene-expression-transcriptional-repression-and-induction.html#lesson).

Additional information about gene expression (http://genome.wellcome.ac.uk/doc_WTD020757.html) is available from the Wellcome Trust.

What is the epigenome?

DNA modifications that do not change the DNA sequence can affect gene activity. Chemical compounds that are added to single genes can regulate their activity; these modifications are known as epigenetic changes. The epigenome comprises all of the chemical compounds that have been added to the entirety of one's DNA (genome) as a way to regulate the activity (expression) of all the genes within the genome. The chemical compounds of the epigenome are not part of the DNA sequence, but are on or attached to DNA ("epi-" means above in Greek). Epigenomic modifications remain as cells divide and in some cases can be inherited through the generations. Environmental influences, such as a person's diet and exposure to pollutants, can also impact the epigenome.

Epigenetic changes can help determine whether genes are turned on or off and can influence the production of proteins in certain cells, ensuring that only necessary proteins are produced. For example, proteins that promote bone growth are not produced in muscle cells. Patterns of epigenome modification vary among individuals, different tissues within an individual, and even different cells.

A common type of epigenomic modification is called methylation. Methylation involves attaching small molecules called methyl groups, each consisting of one carbon atom and three hydrogen atoms, to segments of DNA. When methyl groups are added to a particular gene, that gene is turned off or silenced, and no protein is produced from that gene.

Because errors in the epigenetic process, such as modifying the wrong gene or failing to add a compound to a gene, can lead to abnormal gene activity or inactivity, they can cause genetic disorders. Conditions including cancers, metabolic disorders, and degenerative disorders have all been found to be related to epigenetic errors.

Scientists continue to explore the relationship between the genome and the chemical compounds that modify it. In particular, they are studying what effect the modifications have on gene function, protein production, and human health.

For more information about the epigenome:

The National Institutes of Health (NIH) offers the NIH Roadmap Epigenomics Project (http://roadmapepigenomics.org/), which provides epigenome maps of a variety of cells to begin to assess the relationship between epigenomics and human disease.

The National Center for Biotechnology Information (NCBI) provides the NCBI Epigenomics (http://www.ncbi.nlm.nih.gov/epigenomics) database of maps of the epigenomes of various species and many cell types.

Human Epigenome Atlas (http://www.genboree.org/epigenomeatlas/index.rhtml) from Baylor College of Medicine allows for comparison of the epigenomes of many species and cell types.

Ongoing research is being done with the Human Epigenome Project (http://www.epigenome.org/).

The University of Utah provides an interactive epigenetics tutorial (http://learn.genetics.utah.edu/content/epigenetics/).

The National Human Genome Research Institute provides a fact sheet (http://www.genome.gov/27532724) on Epigenomics.

Many tools for understanding epigenomics are available through the NIH Common Fund Epigenomics Project (https://commonfund.nih.gov/epigenomics/).

How do cells divide?

There are two types of cell division: mitosis and meiosis. Most of the time when people refer to "cell division," they mean mitosis, the process of making new body cells. Meiosis is the type of cell division that creates egg and sperm cells.

Mitosis is a fundamental process for life. During mitosis, a cell duplicates all of its contents, including its chromosomes, and splits to form two identical daughter cells. Because this process is so critical, the steps of mitosis are carefully controlled by a number of genes. When mitosis is not regulated correctly, health problems such as cancer can result.

The other type of cell division, meiosis, ensures that humans have the same number of chromosomes in each generation. It is a two-step process that reduces the chromosome number by half—from 46 to 23—to form sperm and egg cells. When the sperm and egg cells unite at conception, each contributes 23 chromosomes so the resulting embryo will have the usual 46. Meiosis also allows genetic variation through a process of DNA shuffling while the cells are dividing.

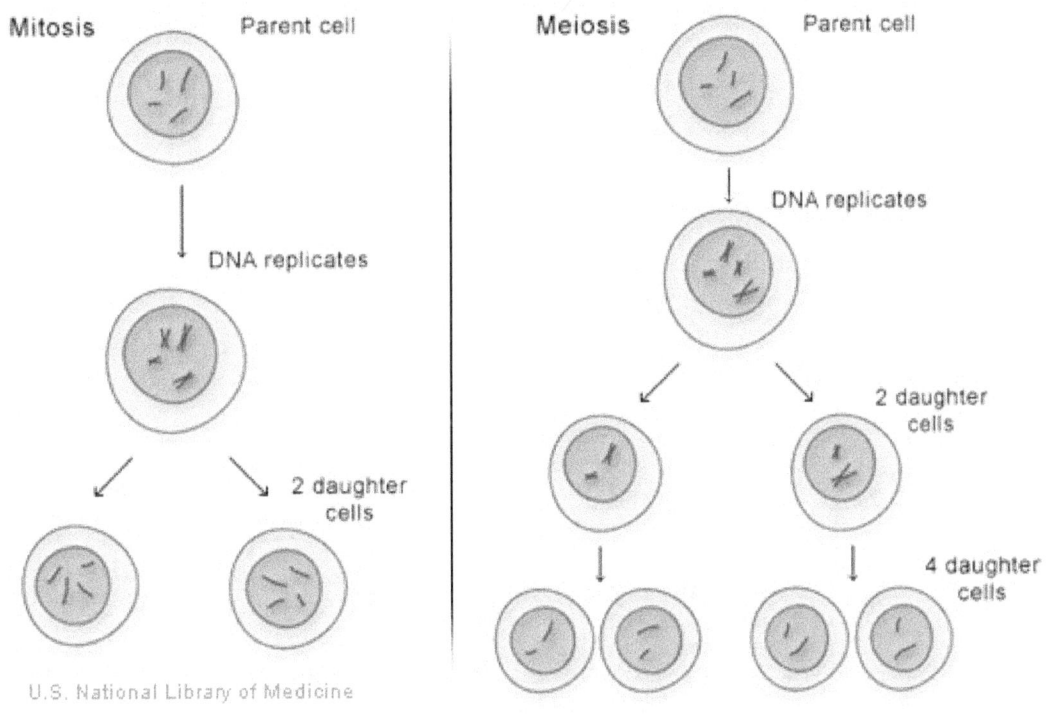

Mitosis and meiosis, the two types of cell division.

For more information about cell division:

Information about mitosis (http://geneed.nlm.nih.gov/topic_subtopic.php?tid=1&sid=2) and meiosis (http://geneed.nlm.nih.gov/topic_subtopic.php?tid=1&sid=3) is available from GeneEd.

The University of Illinois at Chicago offers an outline of meiosis, mitosis, and regulation of the cell cycle (http://www.uic.edu/classes/bios/bios100/lecturesf04am/lect16.htm).

North Dakota State University's Virtual Cell Animation Collection offers videos that illustrate the processes of mitosis (http://vcell.ndsu.nodak.edu/animations/mitosis/movie-flash.htm) and meiosis (http://vcell.ndsu.nodak.edu/animations/meiosis/movie-flash.htm).

Nature Education's Scitable explains how DNA is copied and distributed during the stages of meiosis (http://www.nature.com/scitable/topicpage/replication-and-distribution-of-dna-during-meiosis-6524853).

How do genes control the growth and division of cells?

A variety of genes are involved in the control of cell growth and division. The cell cycle is the cell's way of replicating itself in an organized, step-by-step fashion. Tight regulation of this process ensures that a dividing cell's DNA is copied properly, any errors in the DNA are repaired, and each daughter cell receives a full set of chromosomes. The cycle has checkpoints (also called restriction points), which allow certain genes to check for mistakes and halt the cycle for repairs if something goes wrong.

If a cell has an error in its DNA that cannot be repaired, it may undergo programmed cell death (apoptosis) (illustration on page 32). Apoptosis is a common process throughout life that helps the body get rid of cells it doesn't need. Cells that undergo apoptosis break apart and are recycled by a type of white blood cell called a macrophage (illustration on page 32). Apoptosis protects the body by removing genetically damaged cells that could lead to cancer, and it plays an important role in the development of the embryo and the maintenance of adult tissues.

Cancer results from a disruption of the normal regulation of the cell cycle. When the cycle proceeds without control, cells can divide without order and accumulate genetic defects that can lead to a cancerous tumor (illustration on page 33).

For more information about cell growth and division:

The National Institutes of Health's Apoptosis Interest Group (http://www.nih.gov/sigs/aig/Aboutapo.html) provides an introduction to programmed cell death.

The National Cancer Institute's fact sheet What is Cancer? (http://www.cancer.gov/cancertopics/cancerlibrary/what-is-cancer) explains the growth of cancerous tumors.

Nature Education's Scitable explans how cells become cancerous (http://www.nature.com/scitable/topicpage/cell-division-and-cancer-14046590).

Illustrations

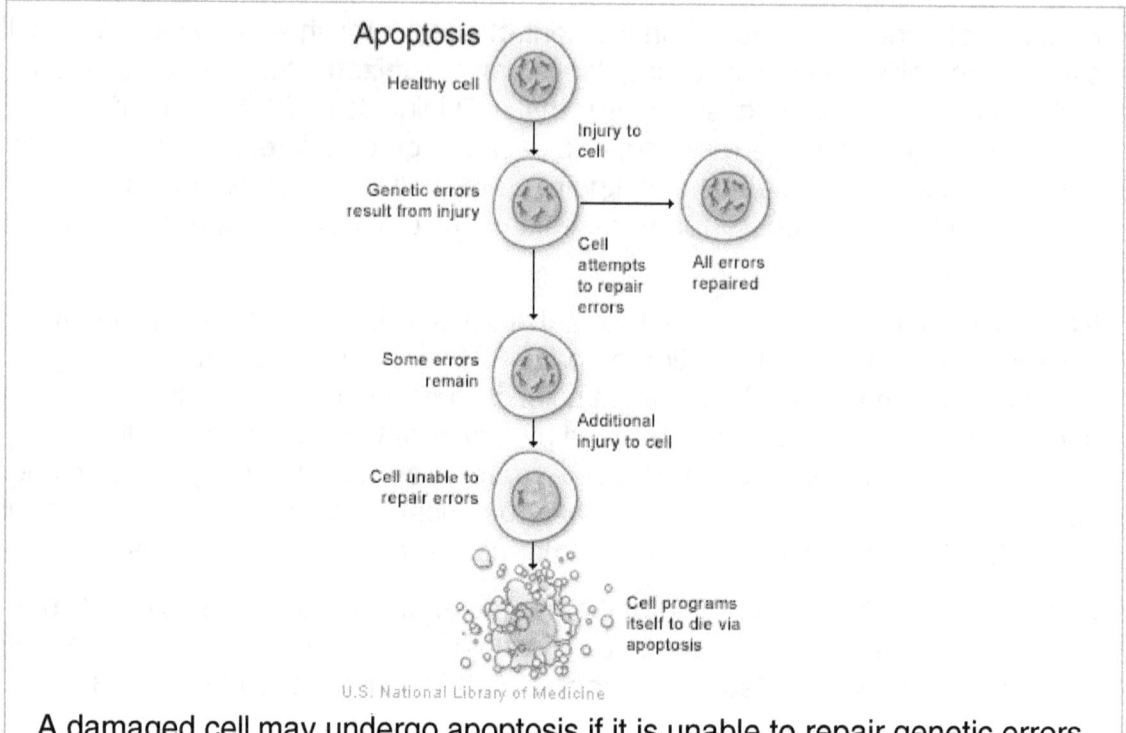

A damaged cell may undergo apoptosis if it is unable to repair genetic errors.

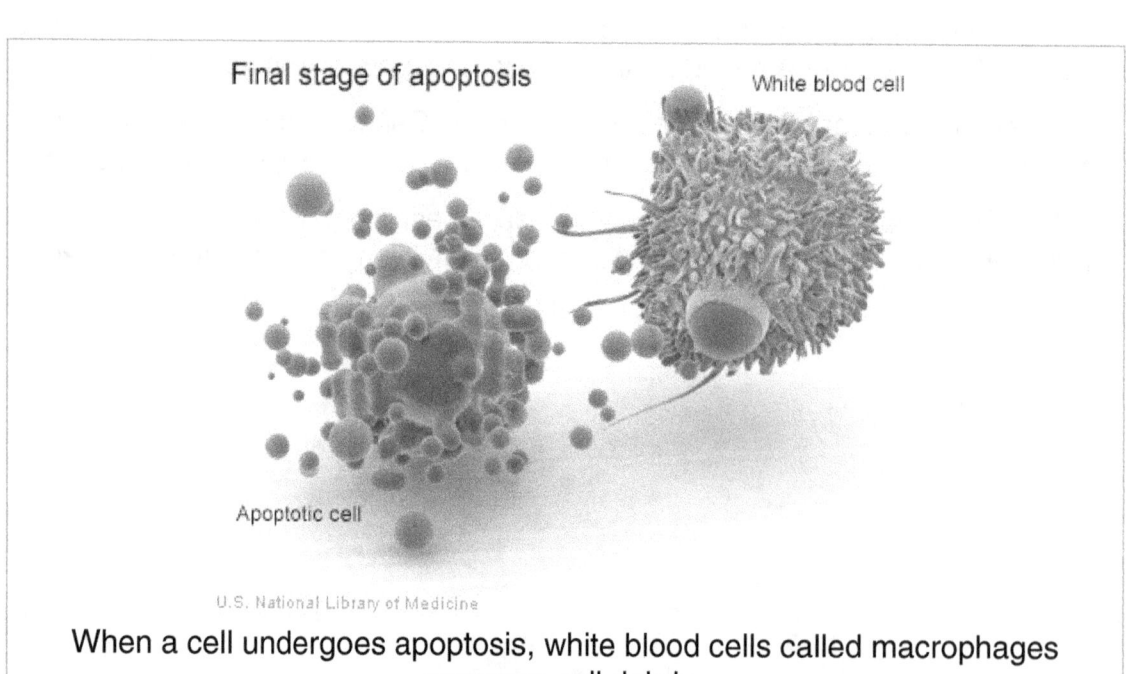

When a cell undergoes apoptosis, white blood cells called macrophages consume cell debris.

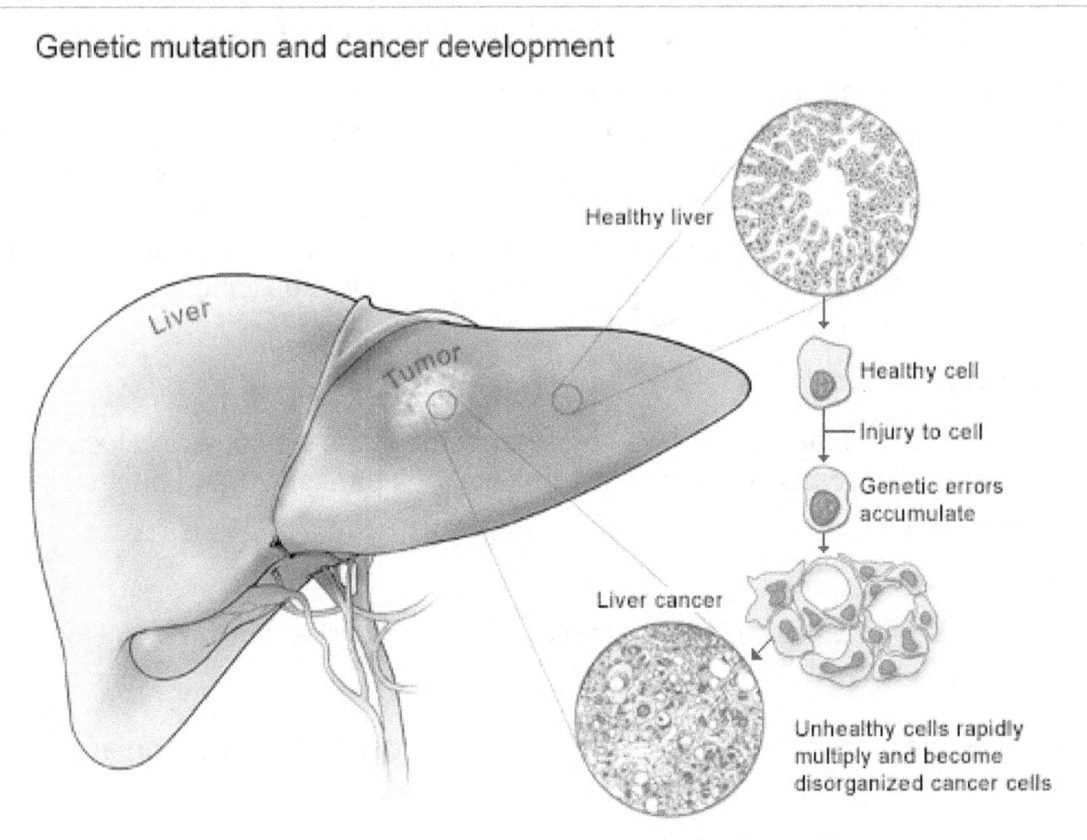

Genetic mutation and cancer development

Healthy liver

Liver

Tumor

Healthy cell

Injury to cell

Genetic errors
accumulate

Liver cancer

Unhealthy cells rapidly
multiply and become
disorganized cancer cells

Cancer results when cells accumulate genetic errors and multiply without control.

How do geneticists indicate the location of a gene?

Geneticists use maps to describe the location of a particular gene on a chromosome. One type of map uses the cytogenetic location to describe a gene's position. The cytogenetic location is based on a distinctive pattern of bands created when chromosomes are stained with certain chemicals. Another type of map uses the molecular location, a precise description of a gene's position on a chromosome. The molecular location is based on the sequence of DNA building blocks (base pairs) that make up the chromosome.

Cytogenetic location

Geneticists use a standardized way of describing a gene's cytogenetic location. In most cases, the location describes the position of a particular band on a stained chromosome:

17q12

It can also be written as a range of bands, if less is known about the exact location:

17q12-q21

The combination of numbers and letters provide a gene's "address" on a chromosome. This address is made up of several parts:

- The chromosome on which the gene can be found. The first number or letter used to describe a gene's location represents the chromosome. Chromosomes 1 through 22 (the autosomes) are designated by their chromosome number. The sex chromosomes are designated by X or Y.

- The arm of the chromosome. Each chromosome is divided into two sections (arms) based on the location of a narrowing (constriction) called the centromere. By convention, the shorter arm is called p, and the longer arm is called q. The chromosome arm is the second part of the gene's address. For example, 5q is the long arm of chromosome 5, and Xp is the short arm of the X chromosome.

- The position of the gene on the p or q arm. The position of a gene is based on a distinctive pattern of light and dark bands that appear when the chromosome is stained in a certain way. The position is usually designated by two digits (representing a region and a band), which are sometimes followed by a decimal point and one or more additional digits (representing sub-bands within a light or dark area). The number indicating the gene position increases with distance from the centromere. For example: 14q21 represents position 21 on the long arm of chromosome 14. 14q21 is closer to the centromere than 14q22.

Sometimes, the abbreviations "cen" or "ter" are also used to describe a gene's cytogenetic location. "Cen" indicates that the gene is very close to the centromere. For example, 16pcen refers to the short arm of chromosome 16 near the centromere. "Ter" stands for terminus, which indicates that the gene is very close to the end of the p or q arm. For example, 14qter refers to the tip of the long arm of chromosome 14. ("Tel" is also sometimes used to describe a gene's location. "Tel" stands for telomeres, which are at the ends of each chromosome. The abbreviations "tel" and "ter" refer to the same location.)

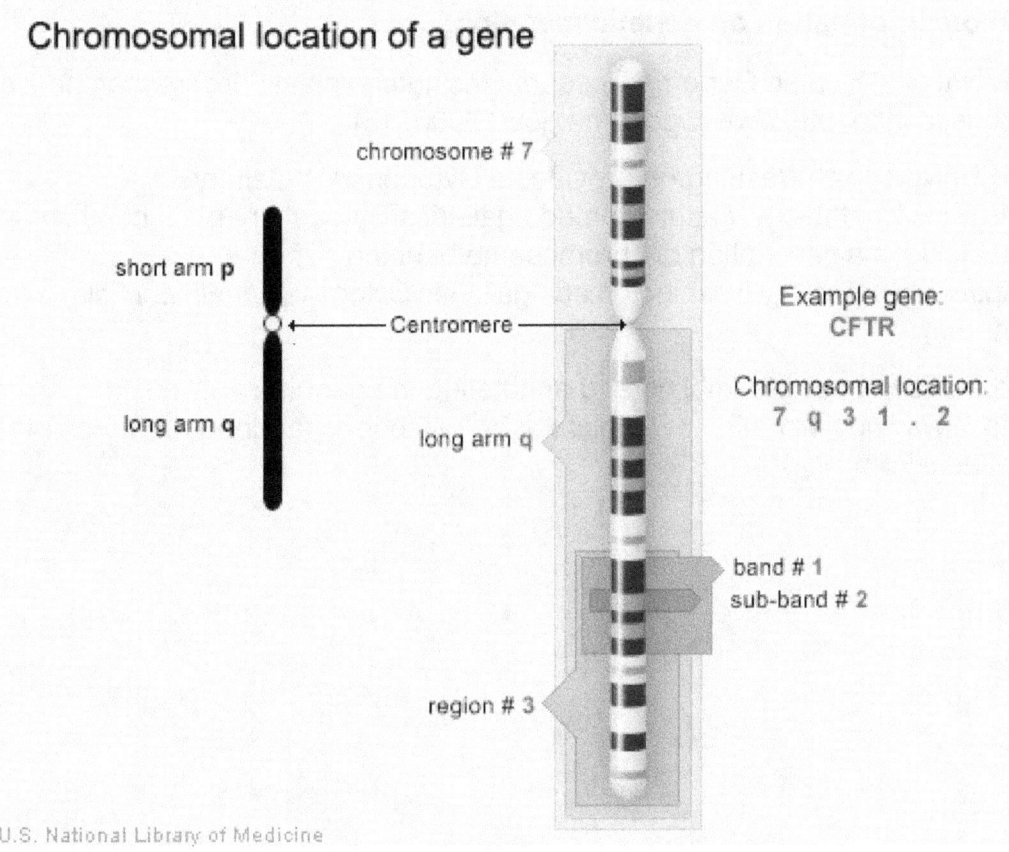

Chromosomal location of a gene

chromosome # 7

short arm p

Centromere

long arm q

long arm q

Example gene:
CFTR

Chromosomal location:
7 q 3 1 . 2

band # 1
sub-band # 2

region # 3

U.S. National Library of Medicine

The CFTR gene is located on the long arm of chromosome 7 at position 7q31.2.

Molecular location

The Human Genome Project, an international research effort completed in 2003, determined the sequence of base pairs for each human chromosome. This sequence information allows researchers to provide a more specific address than the cytogenetic location for many genes. A gene's molecular address pinpoints the location of that gene in terms of base pairs. It describes the gene's precise position on a chromosome and indicates the size of the gene. Knowing the molecular location also allows researchers to determine exactly how far a gene is from other genes on the same chromosome.

Different groups of researchers often present slightly different values for a gene's molecular location. Researchers interpret the sequence of the human genome using a variety of methods, which can result in small differences in a gene's molecular address. Genetics Home Reference presents data from NCBI (http://www.ncbi.nlm.nih.gov/sites/entrez?db=gene) for the molecular location of genes.

For more information on genetic mapping:

The National Human Genome Research Institute explains how researchers create a genetic map (http://www.genome.gov/10000715).

The University of Washington provides a Cytogenetics Gallery (http://www.pathology.washington.edu/galleries/Cytogallery/main.php?file=intro) that includes a description of chromosome banding patterns (http://www.pathology.washington.edu/galleries/Cytogallery/main.php?file=banding+patterns).

Information about assembling and annotating the genome (http://www.ncbi.nlm.nih.gov/bookshelf/br.fcgi?book=handbook&part=ch14) is available from NCBI.

What are gene families?

A gene family is a group of genes that share important characteristics. In many cases, genes in a family share a similar sequence of DNA building blocks (nucleotides). These genes provide instructions for making products (such as proteins) that have a similar structure or function. In other cases, dissimilar genes are grouped together in a family because proteins produced from these genes work together as a unit or participate in the same process.

Classifying individual genes into families helps researchers describe how genes are related to each other. Researchers can use gene families to predict the function of newly identified genes based on their similarity to known genes. Similarities among genes in a family can also be used to predict where and when a specific gene is active (expressed). Additionally, gene families may provide clues for identifying genes that are involved in particular diseases.

Sometimes not enough is known about a gene to assign it to an established family. In other cases, genes may fit into more than one family. No formal guidelines define the criteria for grouping genes together. Classification systems for genes continue to evolve as scientists learn more about the structure and function of genes and the relationships between them.

For more information about gene families

Genetics Home Reference provides information about gene families (http://ghr.nlm.nih.gov/geneFamily) including a brief description of each gene family and a list of the genes included in the family.

The HUGO Gene Nomenclature Committee (http://www.genenames.org/genefamily.html) (HGNC) has classified many human genes into families. Each grouping is given a name and symbol, and contains a table of the genes in that family.

The textbook Human Molecular Genetics (second edition, 1999) provides background information on human gene families (http://www.ncbi.nlm.nih.gov/books/NBK7587/#A663).

The Gene Ontology (http://www.geneontology.org/) database lists the protein products of genes by their location within the cell (cellular component), biological process, and molecular function.

The Reactome (http://www.reactome.org/) database classifies the protein products of genes based on their participation in specific biological pathways. For example, this resource provides tables of genes involved in controlled cell death (apoptosis), cell division, and DNA repair.

Chapter 3
Mutations and Health

Table of Contents

What is a gene mutation and how do mutations occur?

A gene mutation is a permanent change in the DNA sequence that makes up a gene. Mutations range in size from a single DNA building block (DNA base) to a large segment of a chromosome.

Gene mutations occur in two ways: they can be inherited from a parent or acquired during a person's lifetime. Mutations that are passed from parent to child are called hereditary mutations or germline mutations (because they are present in the egg and sperm cells, which are also called germ cells). This type of mutation is present throughout a person's life in virtually every cell in the body.

Mutations that occur only in an egg or sperm cell, or those that occur just after fertilization, are called new (de novo) mutations. De novo mutations may explain genetic disorders in which an affected child has a mutation in every cell, but has no family history of the disorder.

Acquired (or somatic) mutations occur in the DNA of individual cells at some time during a person's life. These changes can be caused by environmental factors such as ultraviolet radiation from the sun, or can occur if a mistake is made as DNA copies itself during cell division. Acquired mutations in somatic cells (cells other than sperm and egg cells) cannot be passed on to the next generation.

Mutations may also occur in a single cell within an early embryo. As all the cells divide during growth and development, the individual will have some cells with the mutation and some cells without the genetic change. This situation is called mosaicism.

Some genetic changes are very rare; others are common in the population. Genetic changes that occur in more than 1 percent of the population are called polymorphisms. They are common enough to be considered a normal variation in the DNA. Polymorphisms are responsible for many of the normal differences between people such as eye color, hair color, and blood type. Although many polymorphisms have no negative effects on a person's health, some of these variations may influence the risk of developing certain disorders.

For more information about mutations:

The Centre for Genetics Education provides a fact sheet discussing changes to the genetic code (http://www.genetics.edu.au/Publications-and-Resources/Genetics-Fact-Sheets/FactSheet4).

More basic information about genetic mutations (http://geneed.nlm.nih.gov/topic_subtopic.php?tid=142&sid=145) is available from GeneEd.

Additional information about genetic changes is available from the University of Utah fact sheet "What is Mutation?" (http://learn.genetics.utah.edu/content/variation/mutation/)

How can gene mutations affect health and development?

To function correctly, each cell depends on thousands of proteins to do their jobs in the right places at the right times. Sometimes, gene mutations prevent one or more of these proteins from working properly. By changing a gene's instructions for making a protein, a mutation can cause the protein to malfunction or to be missing entirely. When a mutation alters a protein that plays a critical role in the body, it can disrupt normal development or cause a medical condition. A condition caused by mutations in one or more genes is called a genetic disorder.

In some cases, gene mutations are so severe that they prevent an embryo from surviving until birth. These changes occur in genes that are essential for development, and often disrupt the development of an embryo in its earliest stages. Because these mutations have very serious effects, they are incompatible with life.

It is important to note that genes themselves do not cause disease—genetic disorders are caused by mutations that make a gene function improperly. For example, when people say that someone has "the cystic fibrosis gene," they are usually referring to a mutated version of the CFTR gene, which causes the disease. All people, including those without cystic fibrosis, have a version of the CFTR gene.

For more information about mutations and genetic disorders:

The Centre for Genetics Education offers a fact sheet about genetic changes that lead to disorders (http://www.genetics.edu.au/Publications-and-Resources/Genetics-Fact-Sheets/FactSheet5).

The Tech Museum of Innovation offers a brief overview of genetic mutations and disease (http://genetics.thetech.org/about-genetics/mutations-and-disease).

Do all gene mutations affect health and development?

No; only a small percentage of mutations cause genetic disorders—most have no impact on health or development. For example, some mutations alter a gene's DNA sequence but do not change the function of the protein made by the gene.

Often, gene mutations that could cause a genetic disorder are repaired by certain enzymes before the gene is expressed and an altered protein is produced. Each cell has a number of pathways through which enzymes recognize and repair mistakes in DNA. Because DNA can be damaged or mutated in many ways, DNA repair is an important process by which the body protects itself from disease.

A very small percentage of all mutations actually have a positive effect. These mutations lead to new versions of proteins that help an individual better adapt to changes in his or her environment. For example, a beneficial mutation could result in a protein that protects an individual and future generations from a new strain of bacteria.

Because a person's genetic code can have a large number of mutations with no effect on health, diagnosing genetic conditions can be difficult. Sometimes, genes thought to be related to a particular genetic condition have mutations, but whether these changes are involved in development of the condition has not been determined; these genetic changes are known as variants of unknown significance (VOUS). Sometimes, no mutations are found in suspected disease-related genes, but mutations are found in other genes whose relationship to a particular genetic condition is unknown. It is difficult to know whether these variants are involved in the disease.

For more information about DNA repair and the health effects of gene mutations:

The University of Utah Genetic Science Learning Center provides information about genetic disorders (http://learn.genetics.utah.edu/content/disorders/) that explains why some mutations cause disorders but others do not.

The National Coalition for Health Professional Education in Genetics explains how mutations can be harmful, neutral, or beneficial (http://www.nchpeg.org/dentistry/index.php?option=com_content&view=article&id=22&Itemid=55&limitstart=3).

What kinds of gene mutations are possible?

The DNA sequence of a gene can be altered in a number of ways. Gene mutations have varying effects on health, depending on where they occur and whether they alter the function of essential proteins. The types of mutations include:

Missense mutation (illustration on page 45)

> This type of mutation is a change in one DNA base pair that results in the substitution of one amino acid for another in the protein made by a gene.

Nonsense mutation (illustration on page 45)

> A nonsense mutation is also a change in one DNA base pair. Instead of substituting one amino acid for another, however, the altered DNA sequence prematurely signals the cell to stop building a protein. This type of mutation results in a shortened protein that may function improperly or not at all.

Insertion (illustration on page 46)

> An insertion changes the number of DNA bases in a gene by adding a piece of DNA. As a result, the protein made by the gene may not function properly.

Deletion (illustration on page 46)

> A deletion changes the number of DNA bases by removing a piece of DNA. Small deletions may remove one or a few base pairs within a gene, while larger deletions can remove an entire gene or several neighboring genes. The deleted DNA may alter the function of the resulting protein(s).

Duplication (illustration on page 47)

> A duplication consists of a piece of DNA that is abnormally copied one or more times. This type of mutation may alter the function of the resulting protein.

Frameshift mutation (illustration on page 47)

> This type of mutation occurs when the addition or loss of DNA bases changes a gene's reading frame. A reading frame consists of groups of 3 bases that each code for one amino acid. A frameshift mutation shifts the grouping of these bases and changes the code for amino acids. The resulting protein is usually nonfunctional. Insertions, deletions, and duplications can all be frameshift mutations.

Repeat expansion (illustration on page 48)

Nucleotide repeats are short DNA sequences that are repeated a number of times in a row. For example, a trinucleotide repeat is made up of 3-base-pair sequences, and a tetranucleotide repeat is made up of 4-base-pair sequences. A repeat expansion is a mutation that increases the number of times that the short DNA sequence is repeated. This type of mutation can cause the resulting protein to function improperly.

For more information about the types of gene mutations:

The National Human Genome Research Institute offers a Talking Glossary of Genetic Terms (http://www.genome.gov/Glossary/). This resource includes definitions, diagrams, and detailed audio descriptions of several of the gene mutations listed above.

A brief explanation of different mutation types (http://www.uvm.edu/~cgep/Education/Mutations.html) is available from the University of Vermont.

Nature Education's Scitable, provides an overview of the different types of mutation that can occur (http://www.nature.com/scitable/topicpage/dna-is-constantly-changing-through-the-process-6524898).

The Education Portal offers detailed information about the different kinds of mutations and their effects:

- How Point Mutations, Insertions, and Deletions Affect DNA (http://education-portal.com/academy/lesson/how-point-mutations-insertions-and-deletions-affect-dna.html#lesson)

- Effects of Mutations on Protein Function: Missense, Nonsense, and Silent Mutations (http://education-portal.com/academy/lesson/mutagens-and-the-effects-of-frameshift-mutations-definitions-and-examples.html#lesson)

- Effects of Frameshift Mutations: Definitions and Examples (http://education-portal.com/academy/lesson/mutagens-and-the-effects-of-frameshift-mutations-definitions-and-examples.html#lesson)

Illustrations

Missense mutation

Original DNA code for an amino acid sequence.

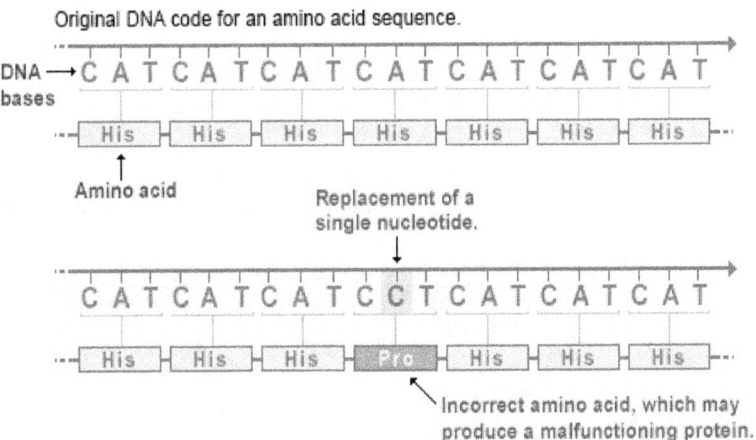

In this example, the nucleotide adenine is replaced by cytosine in the genetic code, introducing an incorrect amino acid into the protein sequence.

Nonsense mutation

Original DNA code for an amino acid sequence.

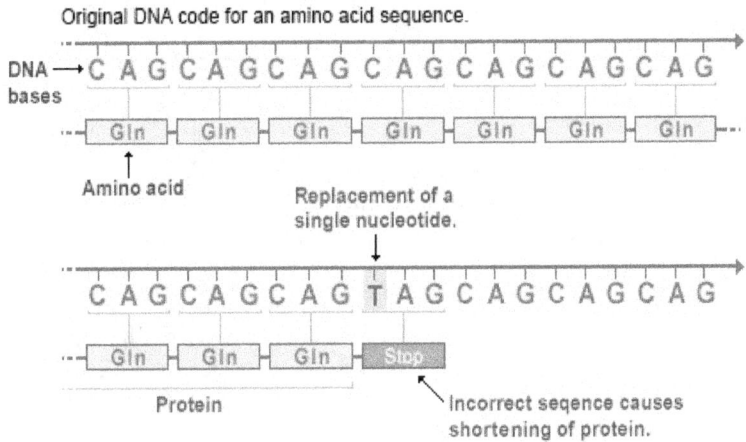

In this example, the nucleotide cytosine is replaced by thymine in the DNA code, signaling the cell to shorten the protein.

Insertion mutation

Original DNA code for an amino acid sequence.

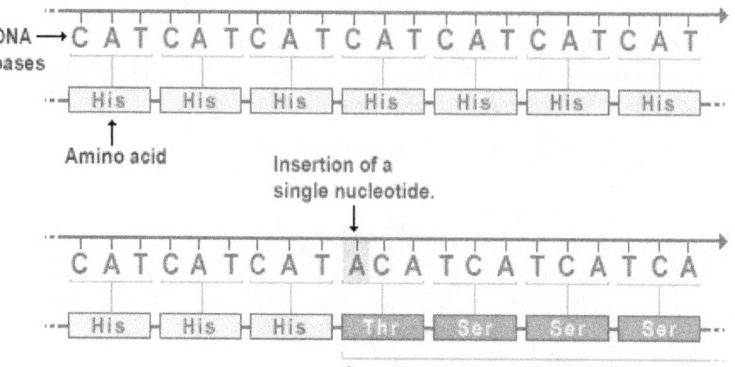

In this example, one nucleotide (adenine) is added in the DNA code, changing the amino acid sequence that follows.

Deletion mutation

Original DNA code for an amino acid sequence.

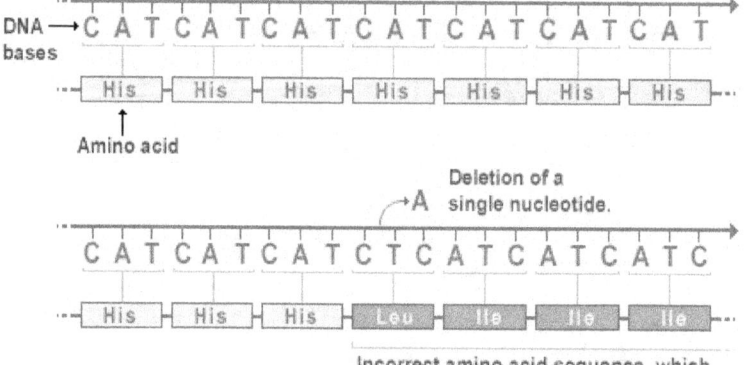

In this example, one nucleotide (adenine) is deleted from the DNA code, changing the amino acid sequence that follows.

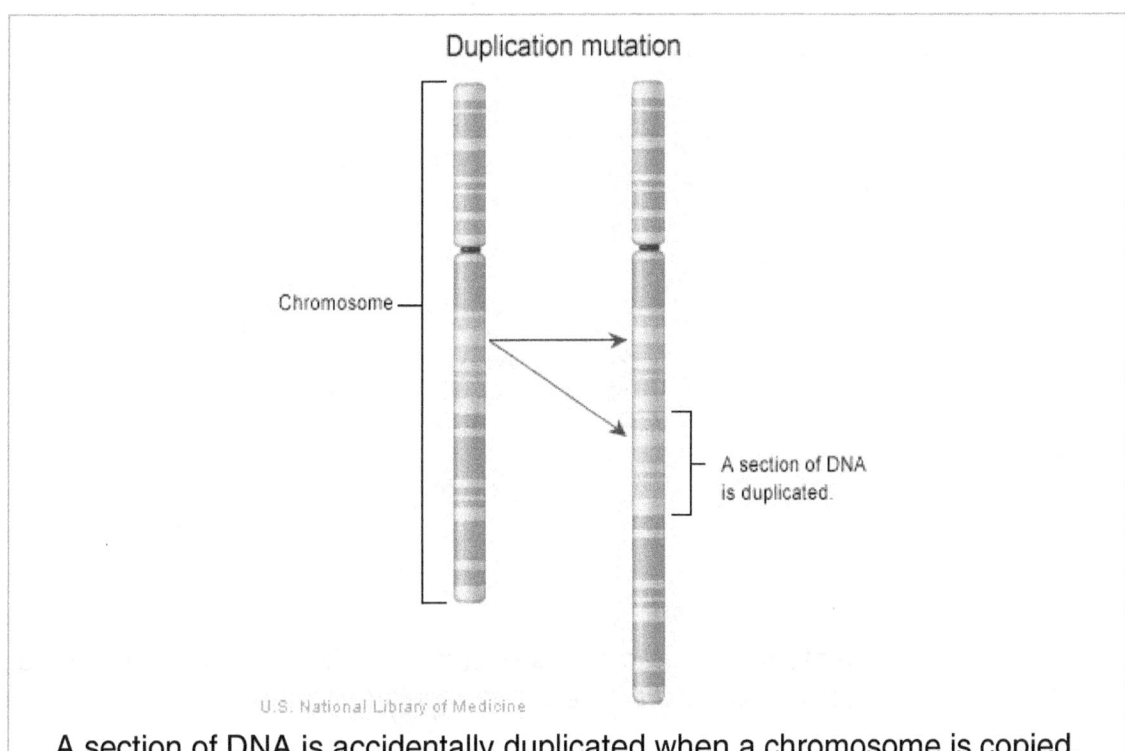

Duplication mutation

Chromosome

A section of DNA is duplicated.

U.S. National Library of Medicine

A section of DNA is accidentally duplicated when a chromosome is copied.

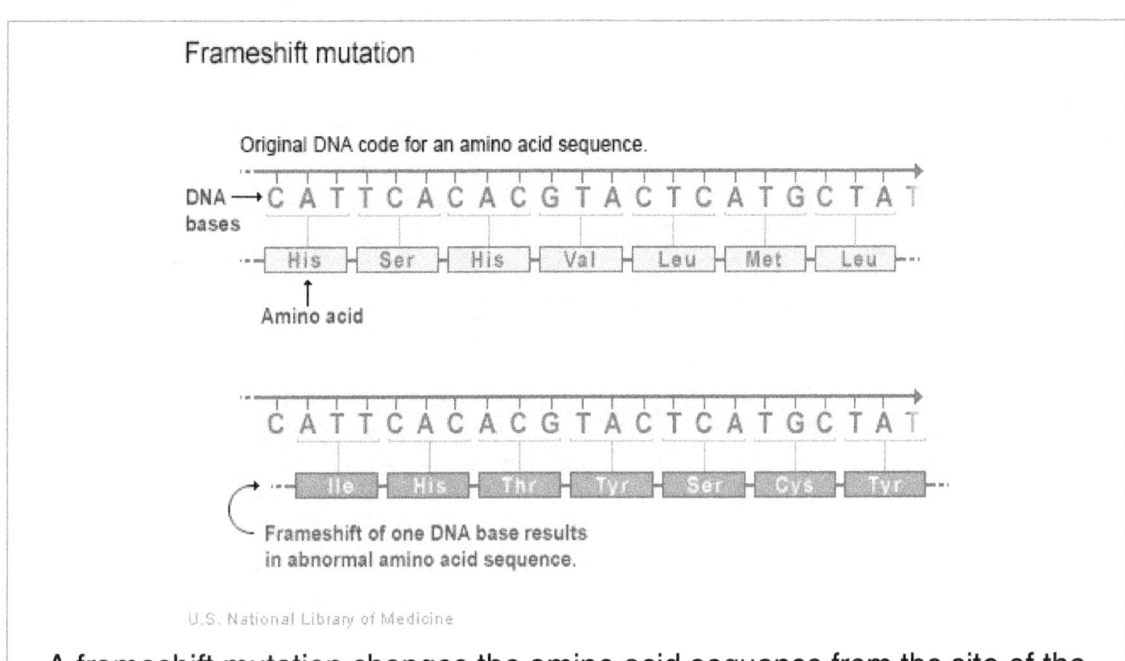

Frameshift mutation

Original DNA code for an amino acid sequence.

DNA bases → C A T T C A C A C G T A C T C A T G C T A T

His — Ser — His — Val — Leu — Met — Leu

Amino acid

C A T T C A C A C G T A C T C A T G C T A T

Ile — His — Thr — Tyr — Ser — Cys — Tyr

Frameshift of one DNA base results in abnormal amino acid sequence.

U.S. National Library of Medicine

A frameshift mutation changes the amino acid sequence from the site of the mutation.

Repeat expansion mutation

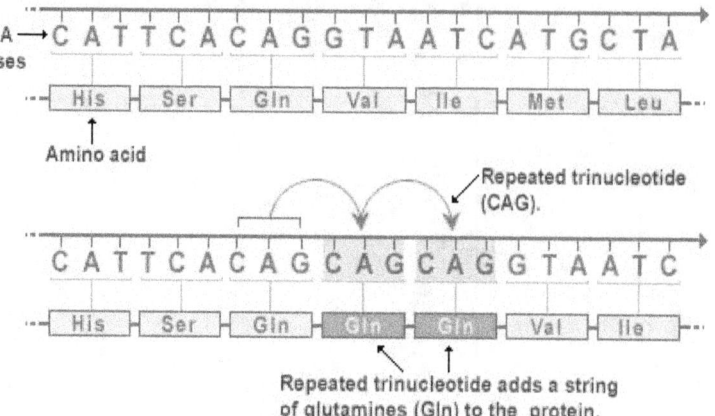

In this example, a repeated trinucleotide sequence (CAG) adds a series of the amino acid glutamine to the resulting protein.

Can a change in the number of genes affect health and development?

People have two copies of most genes, one copy inherited from each parent. In some cases, however, the number of copies varies—meaning that a person can be born with one, three, or more copies of particular genes. Less commonly, one or more genes may be entirely missing. This type of genetic difference is known as copy number variation (CNV).

Copy number variation results from insertions, deletions, and duplications of large segments of DNA. These segments are big enough to include whole genes. Variation in gene copy number can influence the activity of genes and ultimately affect many body functions.

Researchers were surprised to learn that copy number variation accounts for a significant amount of genetic difference between people. More than 10 percent of human DNA appears to contain these differences in gene copy number. While much of this variation does not affect health or development, some differences likely influence a person's risk of disease and response to certain drugs. Future research will focus on the consequences of copy number variation in different parts of the genome and study the contribution of these variations to many types of disease.

For more information about copy number variation:

The Howard Hughes Medical Institute discusses the results of recent research on copy number variation in the news release, Genetic Variation: We're More Different Than We Thought (http://www.hhmi.org/news/scherer20061123.html).

More information about copy number variation (http://www.dnalc.org/view/552-Copy-Number-Variants.html) is available in a video from Cold Spring Harbor Laboratory.

For people interested in more technical data, several institutions provide databases of structural differences in human DNA, including copy number variation:

- Database of Genomic Variants (http://dgv.tcag.ca/dgv/app/home)
- The Sanger Institute: Database of Chromosomal Imbalance and Phenotype in Humans using Ensembl Resources (DECIPHER (http://decipher.sanger.ac.uk/))

Can changes in the number of chromosomes affect health and development?

Human cells normally contain 23 pairs of chromosomes, for a total of 46 chromosomes in each cell (illustration on page 52). A change in the number of chromosomes can cause problems with growth, development, and function of the body's systems. These changes can occur during the formation of reproductive cells (eggs and sperm), in early fetal development, or in any cell after birth. A gain or loss of chromosomes from the normal 46 is called aneuploidy.

A common form of aneuploidy is trisomy, or the presence of an extra chromosome in cells. "Tri-" is Greek for "three"; people with trisomy have three copies of a particular chromosome in cells instead of the normal two copies. Down syndrome is an example of a condition caused by trisomy (illustration on page 53). People with Down syndrome typically have three copies of chromosome 21 in each cell, for a total of 47 chromosomes per cell.

Monosomy, or the loss of one chromosome in cells, is another kind of aneuploidy. "Mono-" is Greek for "one"; people with monosomy have one copy of a particular chromosome in cells instead of the normal two copies. Turner syndrome is a condition caused by monosomy (illustration on page 54). Women with Turner syndrome usually have only one copy of the X chromosome in every cell, for a total of 45 chromosomes per cell.

Rarely, some cells end up with complete extra sets of chromosomes. Cells with one additional set of chromosomes, for a total of 69 chromosomes, are called triploid (illustration on page 55). Cells with two additional sets of chromosomes, for a total of 92 chromosomes, are called tetraploid. A condition in which every cell in the body has an extra set of chromosomes is not compatible with life.

In some cases, a change in the number of chromosomes occurs only in certain cells. When an individual has two or more cell populations with a different chromosomal makeup, this situation is called chromosomal mosaicism (illustration on page 56). Chromosomal mosaicism occurs from an error in cell division in cells other than eggs and sperm. Most commonly, some cells end up with one extra or missing chromosome (for a total of 45 or 47 chromosomes per cell), while other cells have the usual 46 chromosomes. Mosaic Turner syndrome is one example of chromosomal mosaicism. In females with this condition, some cells have 45 chromosomes because they are missing one copy of the X chromosome, while other cells have the usual number of chromosomes.

Many cancer cells also have changes in their number of chromosomes. These changes are not inherited; they occur in somatic cells (cells other than eggs or sperm) during the formation or progression of a cancerous tumor.

For more information about chromosomal disorders:

A discussion of how chromosomal abnormalities happen (http://www.genome.gov/11508982#6) is provided by the National Human Genome Research Institute.

The Centre for Genetics Education offers a fact sheet about changes in chromosome number or size (http://www.genetics.edu.au/Publications-and-Resources/Genetics-Fact-Sheets/FactSheet6).

Information about chromosomal changes (http://www.eurogentest.org/index.php?id=611), including changes in the number of chromosomes, is available from EuroGentest.

The University of Leicester's Virtual Genetics Education Center provides an explanation of numerical chromosome aberrations (http://www2.le.ac.uk/departments/genetics/vgec/healthprof/topics/patterns-of-inheritance/chromosomal-abnormalities#numerical-aberrations).

The National Organization for Rare Disorders offers an overview of triploid syndrome (https://www.rarediseases.org/rare-disease-information/rare-diseases/byID/710/viewAbstract).

Chromosomal Mosaicism (http://mosaicism.cfri.ca/index.htm), a web site provided by the University of British Columbia, offers detailed information about mosaic chromosomal abnormalities.

MedlinePlus offers an encyclopedia article about chromosomal mosaicism (http://www.nlm.nih.gov/medlineplus/ency/article/001317.htm).

Illustrations

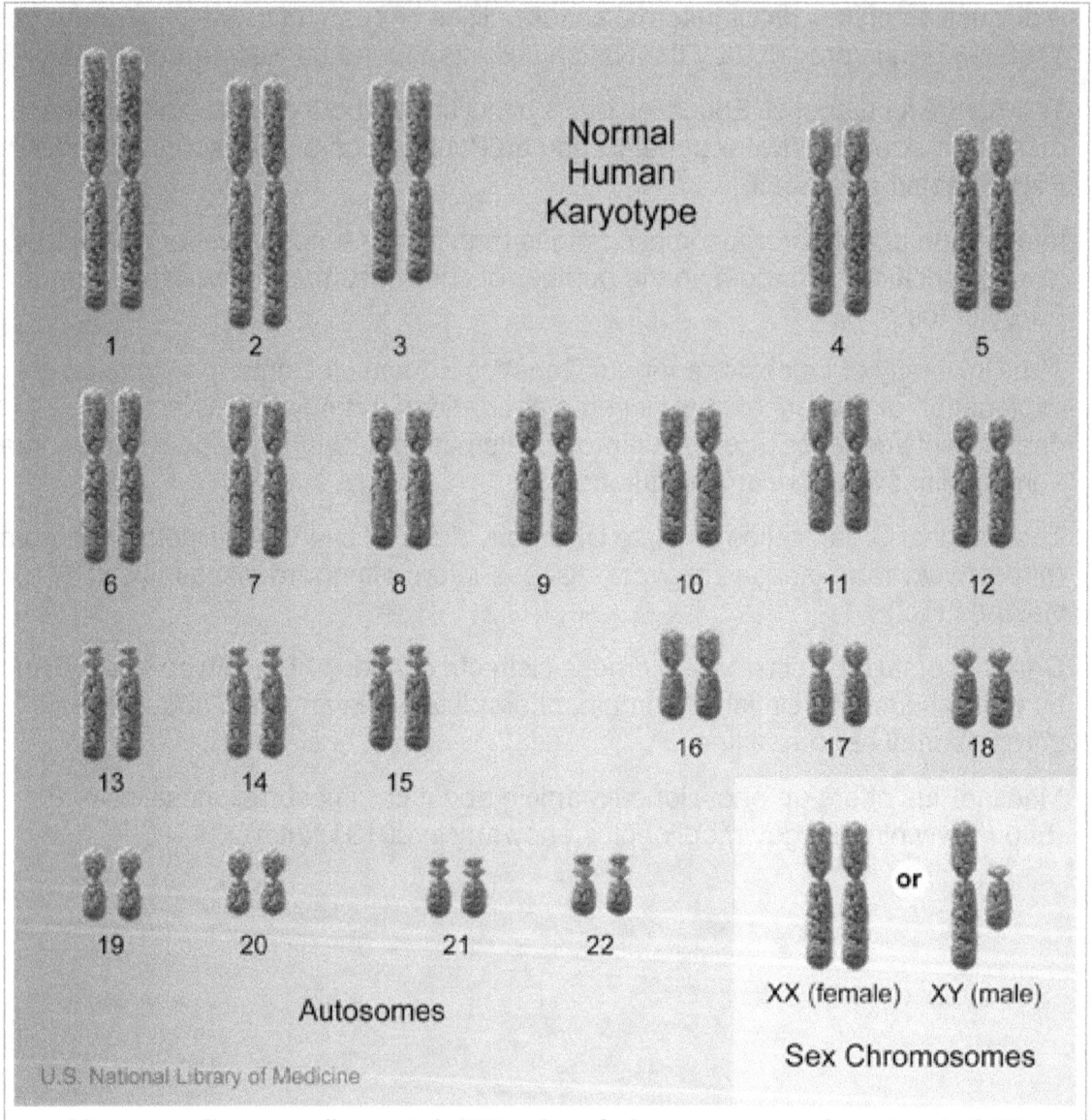

Normal Human Karyotype

Autosomes

Sex Chromosomes

XX (female)   XY (male)

or

U.S. National Library of Medicine

Human cells normally contain 23 pairs of chromosomes, for a total of 46 chromosomes in each cell.

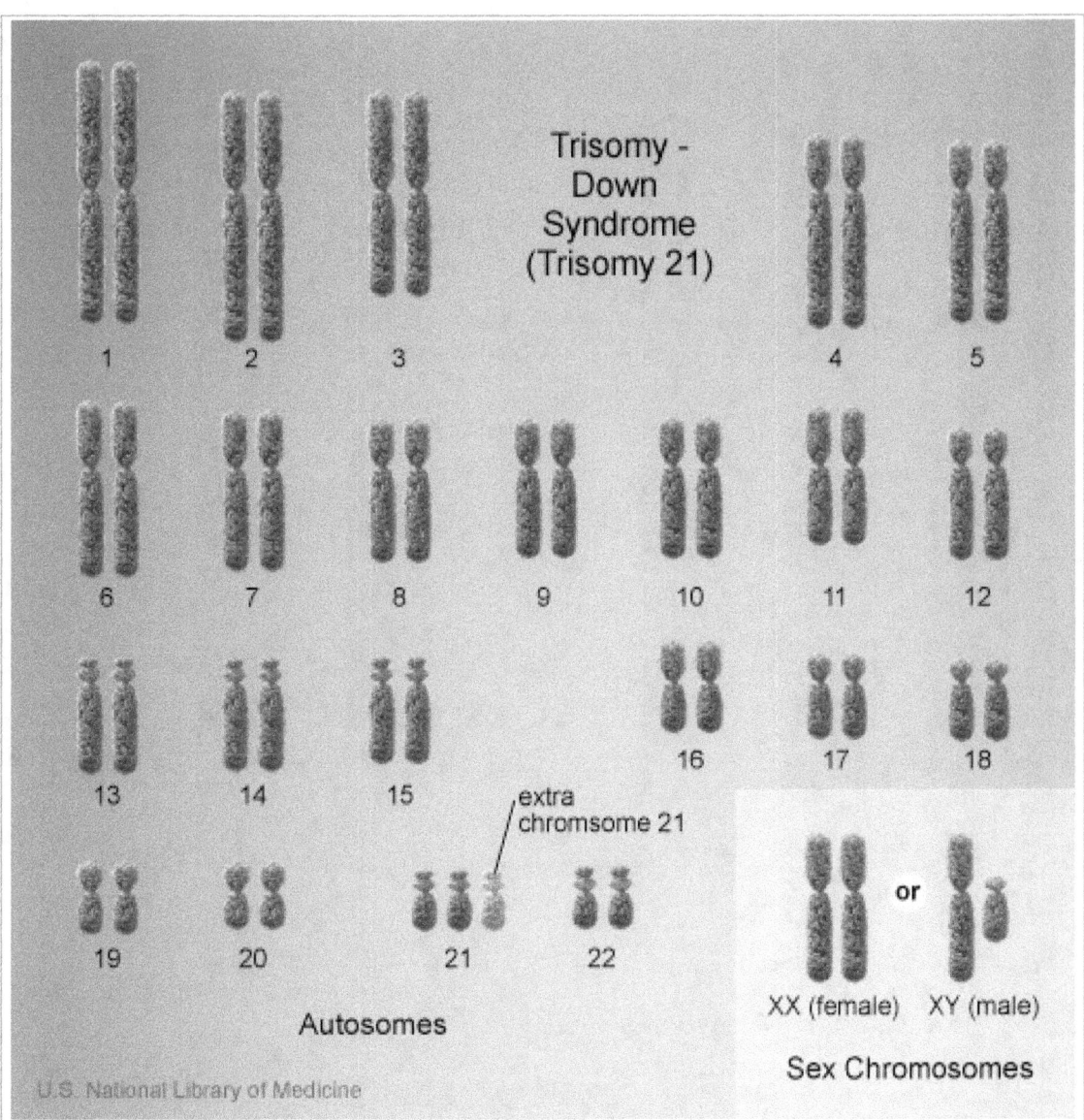

Trisomy - Down Syndrome (Trisomy 21)

1   2   3   4   5

6   7   8   9   10   11   12

13   14   15   16   17   18

extra chromsome 21

19   20   21   22

Autosomes

XX (female)   or   XY (male)

Sex Chromosomes

U.S. National Library of Medicine

Trisomy is the presence of an extra chromosome in cells. Down syndrome is an example of a condition caused by trisomy.

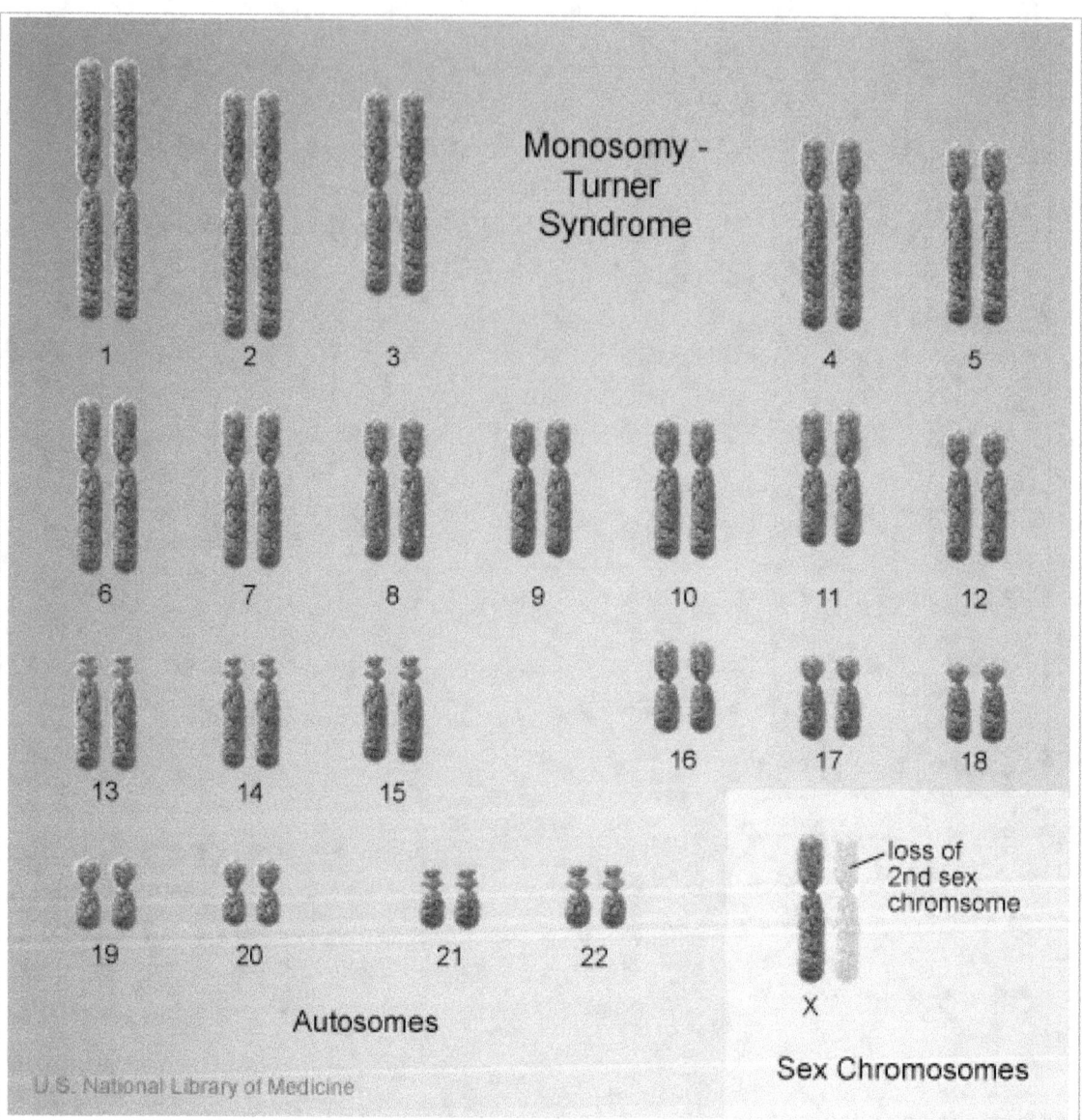

Monosomy is the loss of one chromosome in cells. Turner syndrome is an example of a condition caused by monosomy.

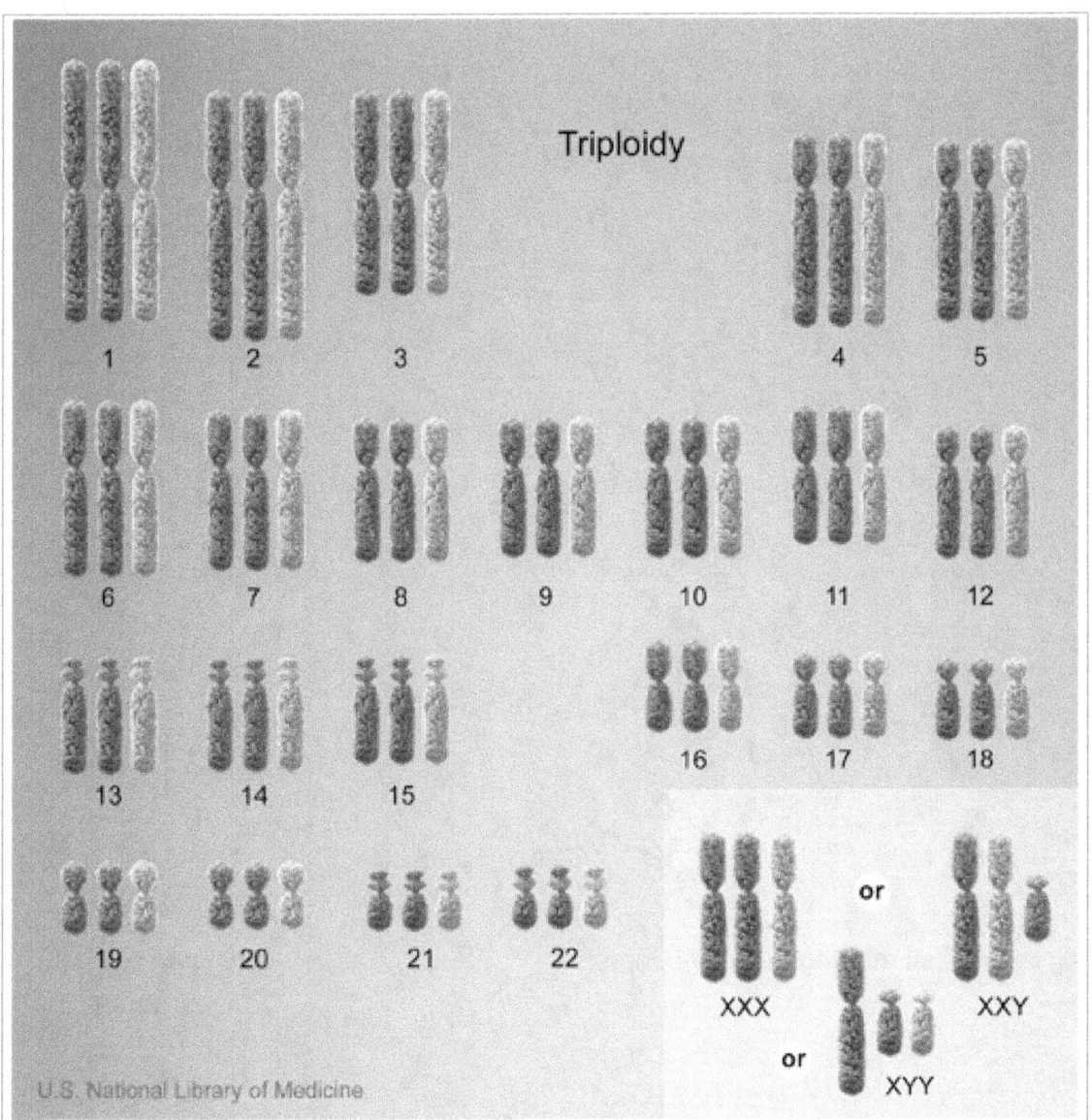

Triploidy

Cells with one additional set of chromosomes, for a total of 69 chromosomes, are called triploid.

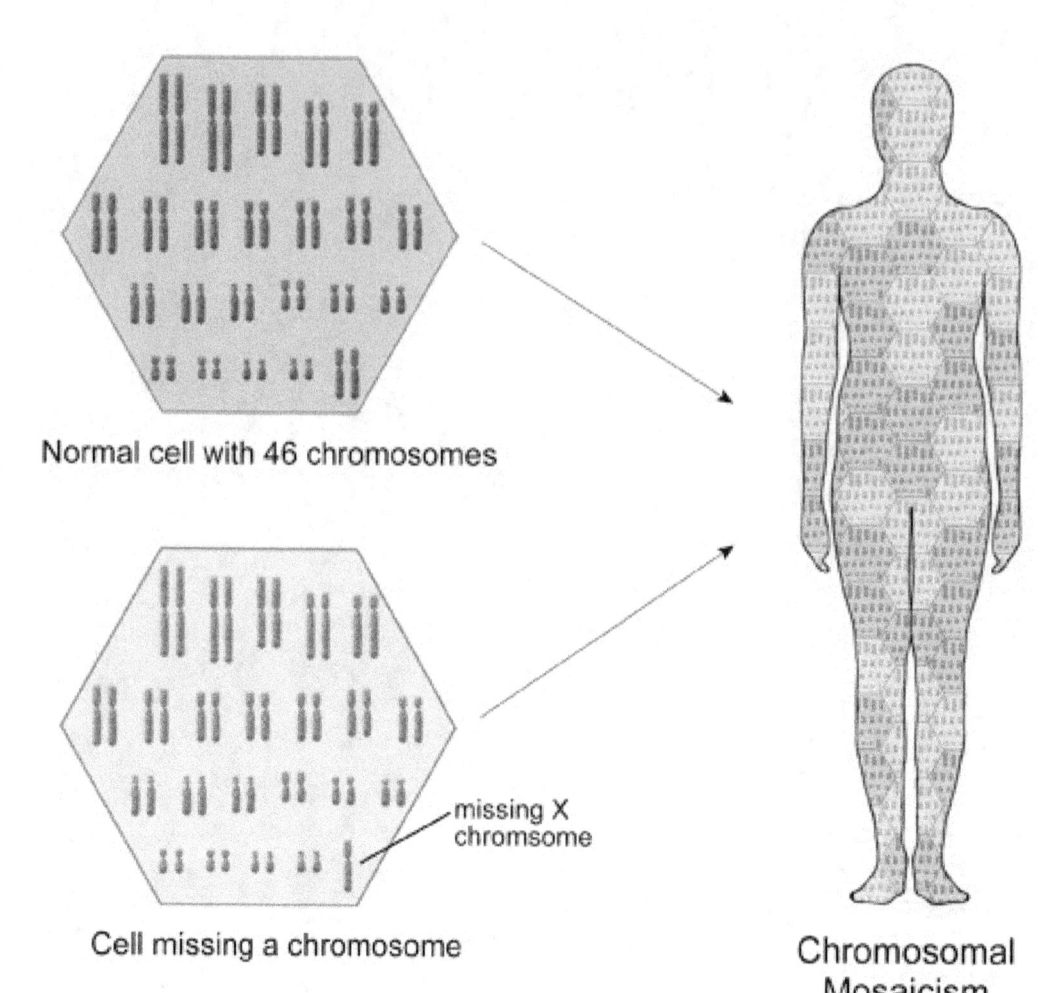

Normal cell with 46 chromosomes

missing X
chromsome

Cell missing a chromosome

Chromosomal
Mosaicism

When an individual has two or more cell populations with a different chromosomal makeup, this situation is called chromosomal mosaicism.

Can changes in the structure of chromosomes affect health and development?

Changes that affect the structure of chromosomes can cause problems with growth, development, and function of the body's systems. These changes can affect many genes along the chromosome and disrupt the proteins made from those genes.

Structural changes can occur during the formation of egg or sperm cells, in early fetal development, or in any cell after birth. Pieces of DNA can be rearranged within one chromosome or transferred between two or more chromosomes. The effects of structural changes depend on their size and location, and whether any genetic material is gained or lost. Some changes cause medical problems, while others may have no effect on a person's health.

Changes in chromosome structure include:

Translocations (illustration: balanced on page 59),
(illustration: unbalanced on page 60)

> A translocation occurs when a piece of one chromosome breaks off and attaches to another chromosome. This type of rearrangement is described as balanced if no genetic material is gained or lost in the cell. If there is a gain or loss of genetic material, the translocation is described as unbalanced.

Deletions (illustration on page 61)

> Deletions occur when a chromosome breaks and some genetic material is lost. Deletions can be large or small, and can occur anywhere along a chromosome.

Duplications (illustration on page 62)

> Duplications occur when part of a chromosome is copied (duplicated) too many times. This type of chromosomal change results in extra copies of genetic material from the duplicated segment.

Inversions (illustration on page 63)

> An inversion involves the breakage of a chromosome in two places; the resulting piece of DNA is reversed and re-inserted into the chromosome. Genetic material may or may not be lost as a result of the chromosome breaks. An inversion that involves the chromosome's constriction point (centromere) is called a pericentric inversion. An inversion that occurs in the long (q) arm or short (p) arm and does not involve the centromere is called a paracentric inversion.

Isochromosomes (illustration on page 64)

An isochromosome is a chromosome with two identical arms. Instead of one long (q) arm and one short (p) arm, an isochromosome has two long arms or two short arms. As a result, these abnormal chromosomes have an extra copy of some genes and are missing copies of other genes.

Dicentric chromosomes (illustration on page 65)

Unlike normal chromosomes, which have a single constriction point (centromere), a dicentric chromosome contains two centromeres. Dicentric chromosomes result from the abnormal fusion of two chromosome pieces, each of which includes a centromere. These structures are unstable and often involve a loss of some genetic material.

Ring chromosomes (illustration on page 66)

Ring chromosomes usually occur when a chromosome breaks in two places and the ends of the chromosome arms fuse together to form a circular structure. The ring may or may not include the chromosome's constriction point (centromere). In many cases, genetic material near the ends of the chromosome is lost.

Many cancer cells also have changes in their chromosome structure. These changes are not inherited; they occur in somatic cells (cells other than eggs or sperm) during the formation or progression of a cancerous tumor.

For more information about structural changes to chromosomes:

The National Human Genome Research Institute provides a list of questions and answers about chromosome abnormalities (http://www.genome.gov/11508982), including a glossary of related terms.

Chromosome Deletion Outreach offers a fact sheet on this topic titled Introduction to Chromosomes (http://chromodisorder.org/Display.aspx?ID=35). This resource includes illustrated explanations of several chromosome abnormalities.

The Centre for Genetics Education provides fact sheets about changes in chromosome number or size (http://www.genetics.edu.au/Publications-and-Resources/Genetics-Fact-Sheets/FactSheet6) and chromosomal rearrangements (translocations) (http://www.genetics.edu.au/Publications-and-Resources/Genetics-Fact-Sheets/FactSheet7).

EuroGentest offers fact sheets about chromosome changes (http://www.eurogentest.org/index.php?id=611) and chromosome translocations (http://www.eurogentest.org/index.php?id=612).

The University of Leicester's Virtual Genetics Education Center provides an explanation of structural chromosome aberrations (http://www2.le.ac.uk/departments/genetics/vgec/healthprof/topics/patterns-of-inheritance/chromosomal-abnormalities#structural-aberrations).

More technical information is available from the textbook Human Molecular Genetics (second edition, 1999) in the section about structural chromosome abnormalities (http://www.ncbi.nlm.nih.gov/books/bv.fcgi?rid=hmg.section.196#209).

The Atlas of Genetics and Cytogenetics in Oncology and Haematology provides a technical introduction to chromosomal aberrations (http://atlasgeneticsoncology.org/Deep/Chromaber.html) and a detailed discussion of ring chromosomes (http://atlasgeneticsoncology.org/Deep/RingChromosID20030.html), particularly their role in cancer.

Illustrations

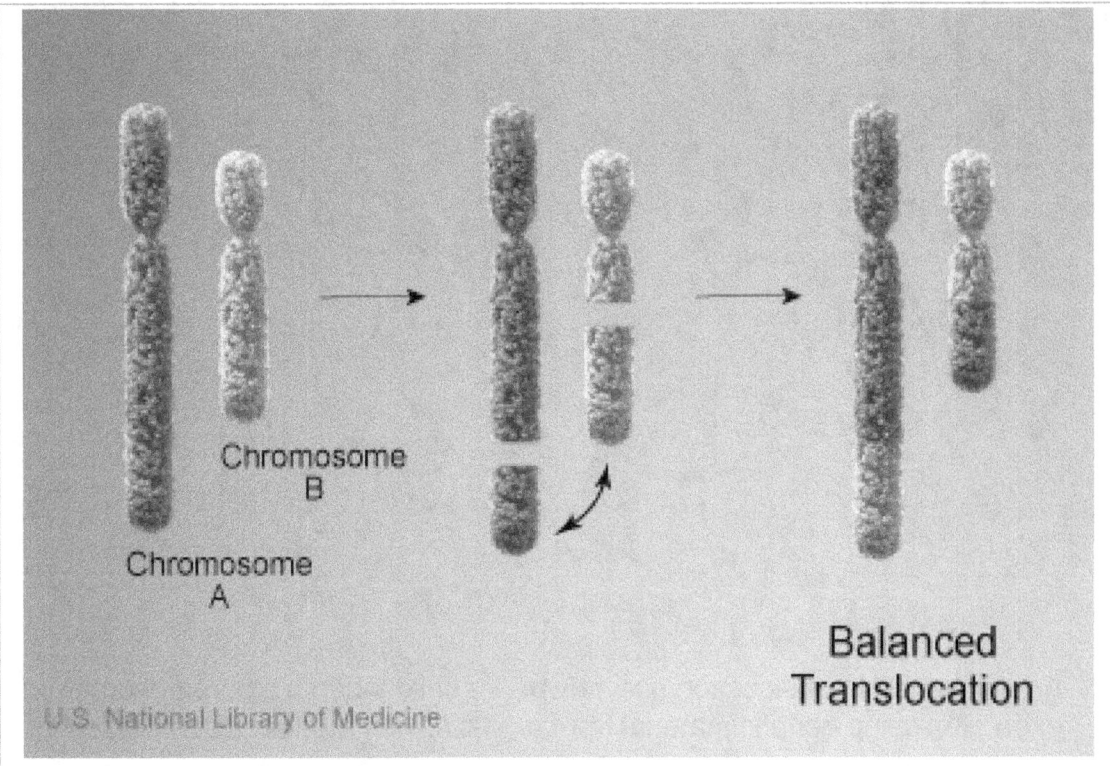

In a balanced translocation, pieces of chromosomes are rearranged but no genetic material is gained or lost in the cell.

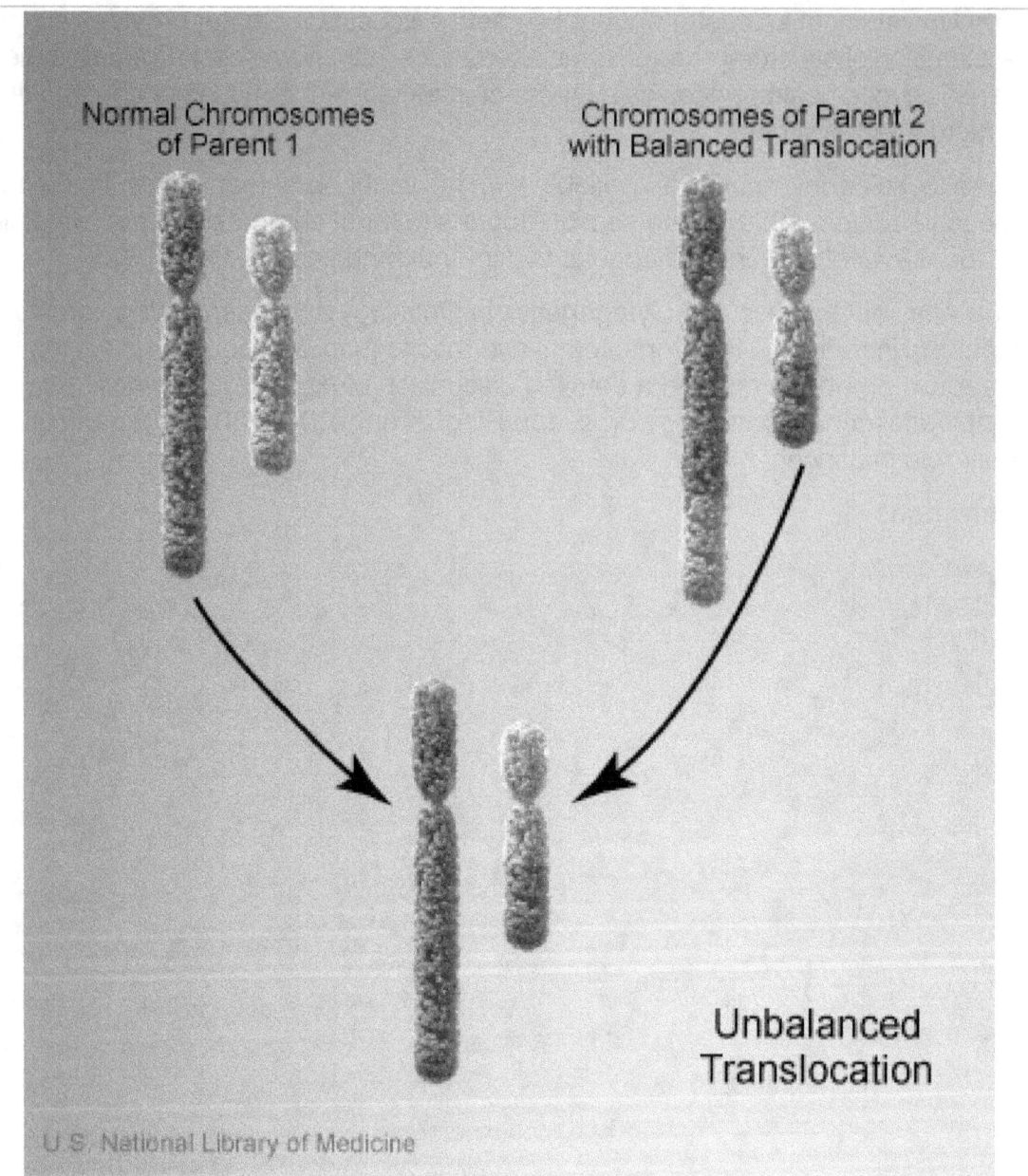

An unbalanced translocation occurs when a child inherits a chromosome with extra or missing genetic material from a parent with a balanced translocation.

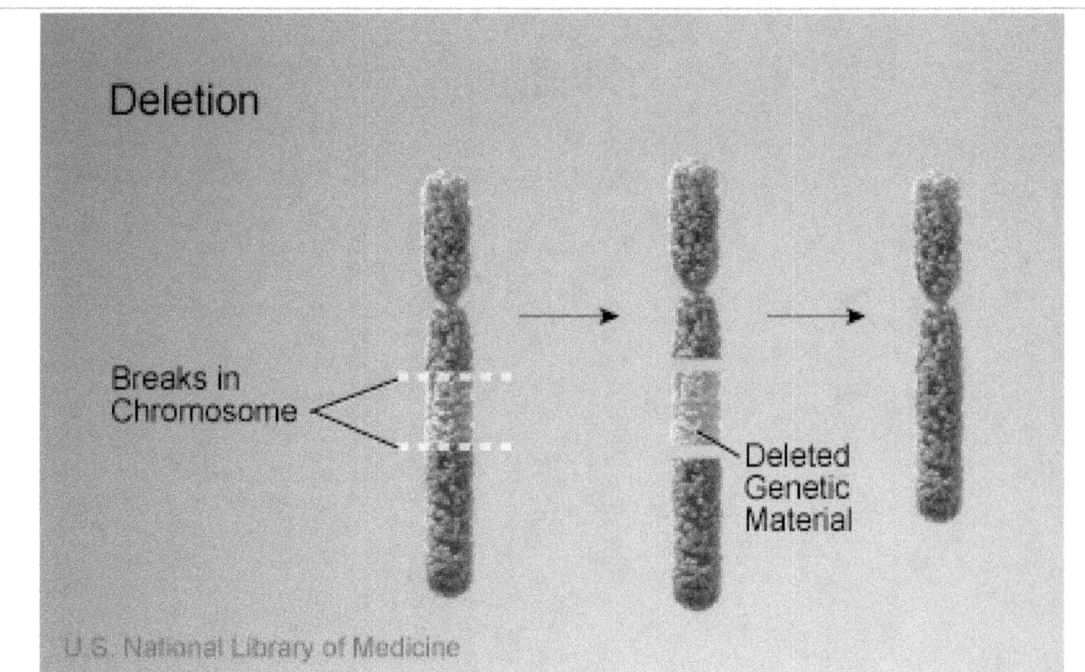

A deletion occurs when a chromosome breaks and some genetic material is lost.

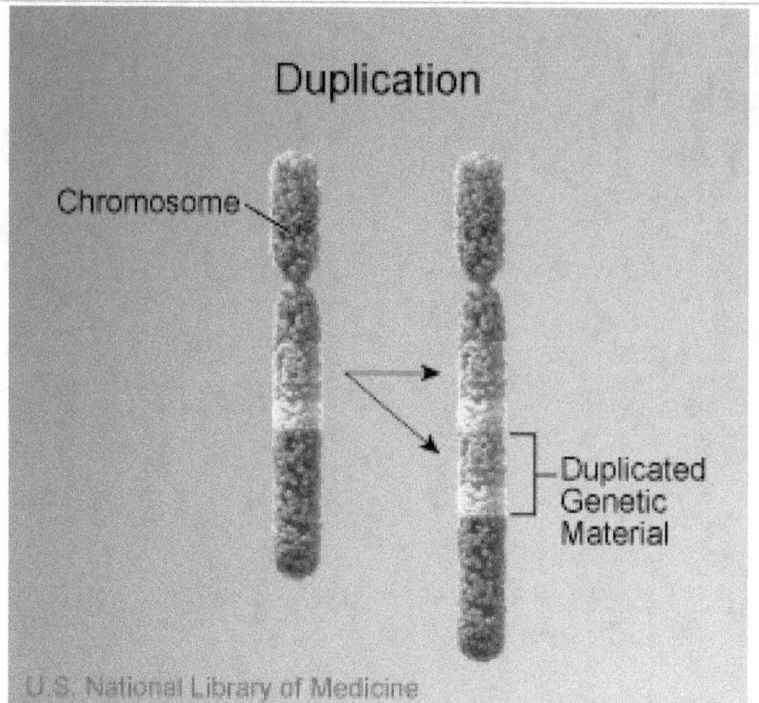

A duplication occurs when part of a chromosome is copied (duplicated) abnormally, resulting in extra genetic material from the duplicated segment.

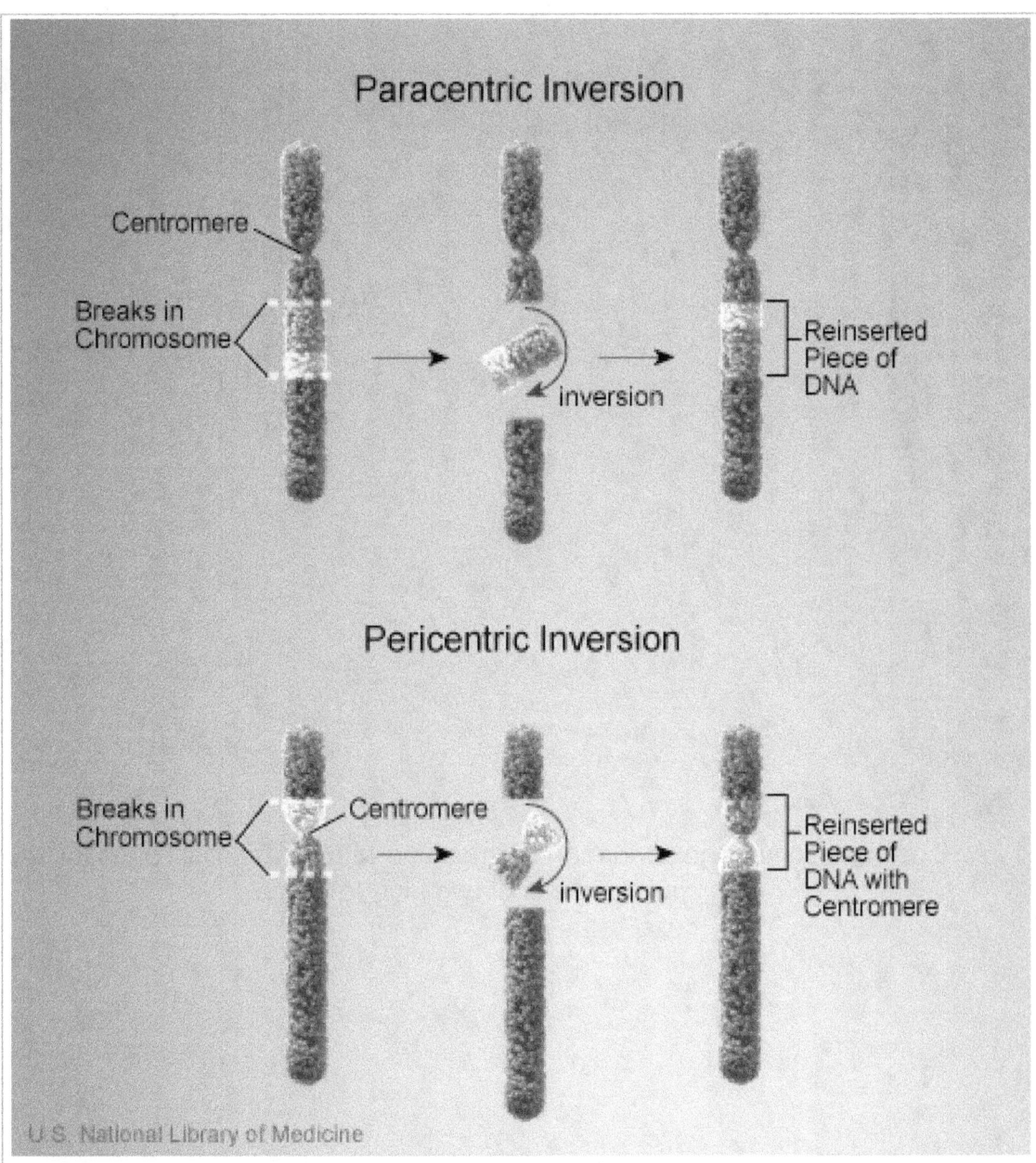

Inversions occur when a chromosome breaks in two places and the resulting piece of DNA is reversed and re-inserted into the chromosome. Inversions that involve the centromere are called pericentric inversions; those that do not involve the centromere are called paracentric inversions.

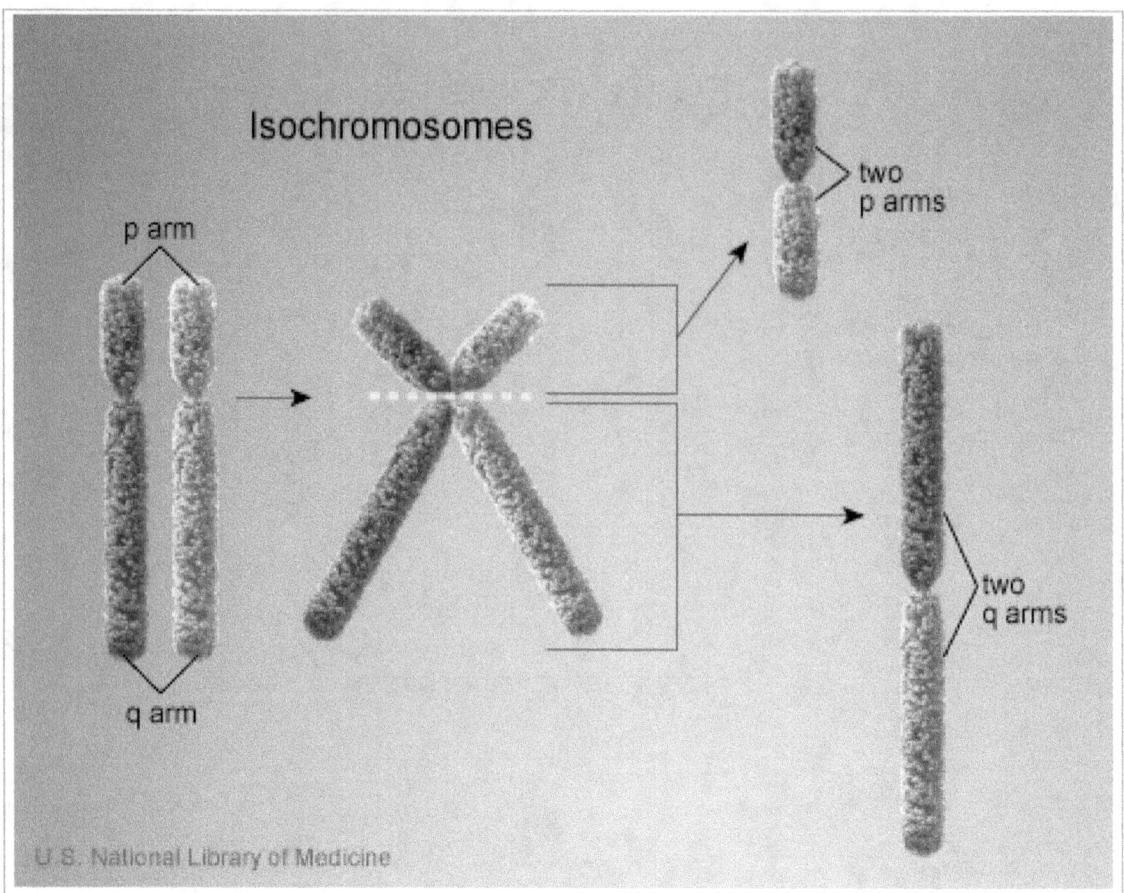

An isochromosome is an abnormal chromosome with two identical arms, either two short (p) arms or two long (q) arms.

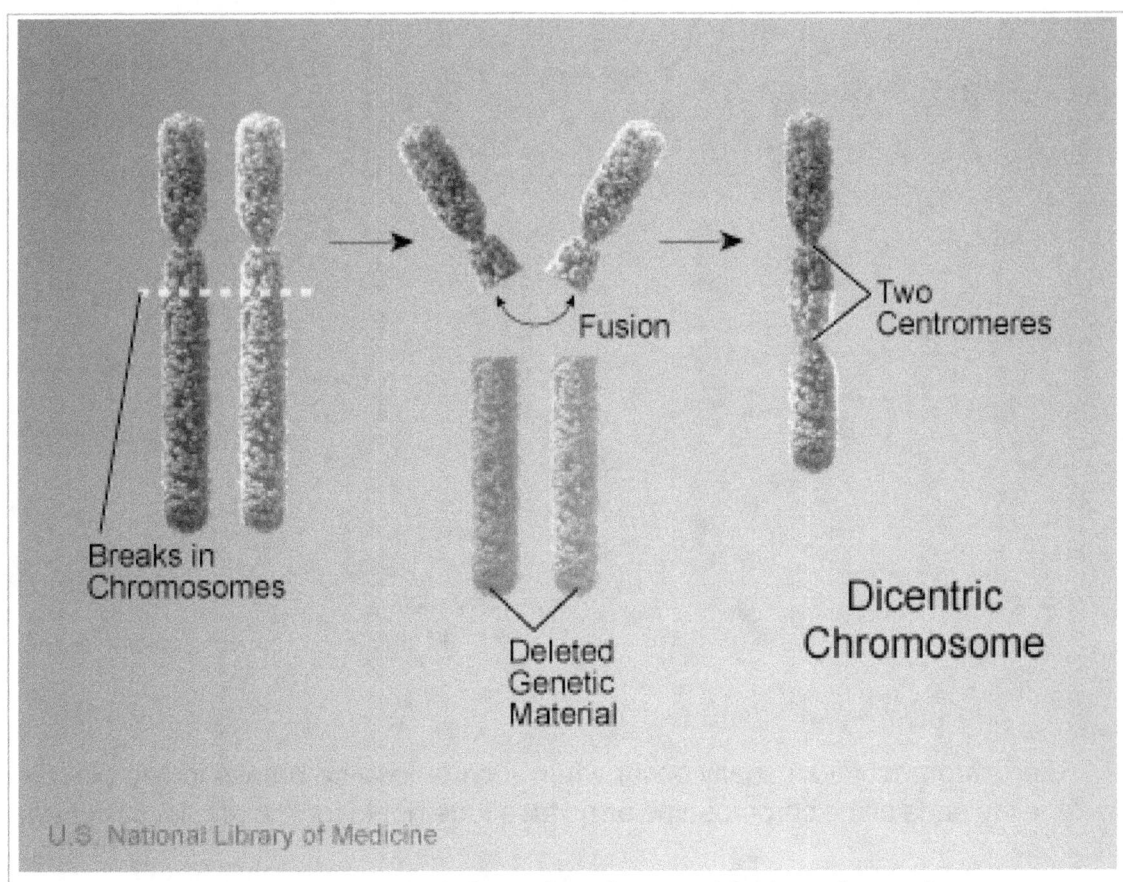

Breaks in
Chromosomes

Fusion

Deleted
Genetic
Material

Two
Centromeres

Dicentric
Chromosome

Dicentric chromosomes result from the abnormal fusion of two chromosome
pieces, each of which includes a centromere.

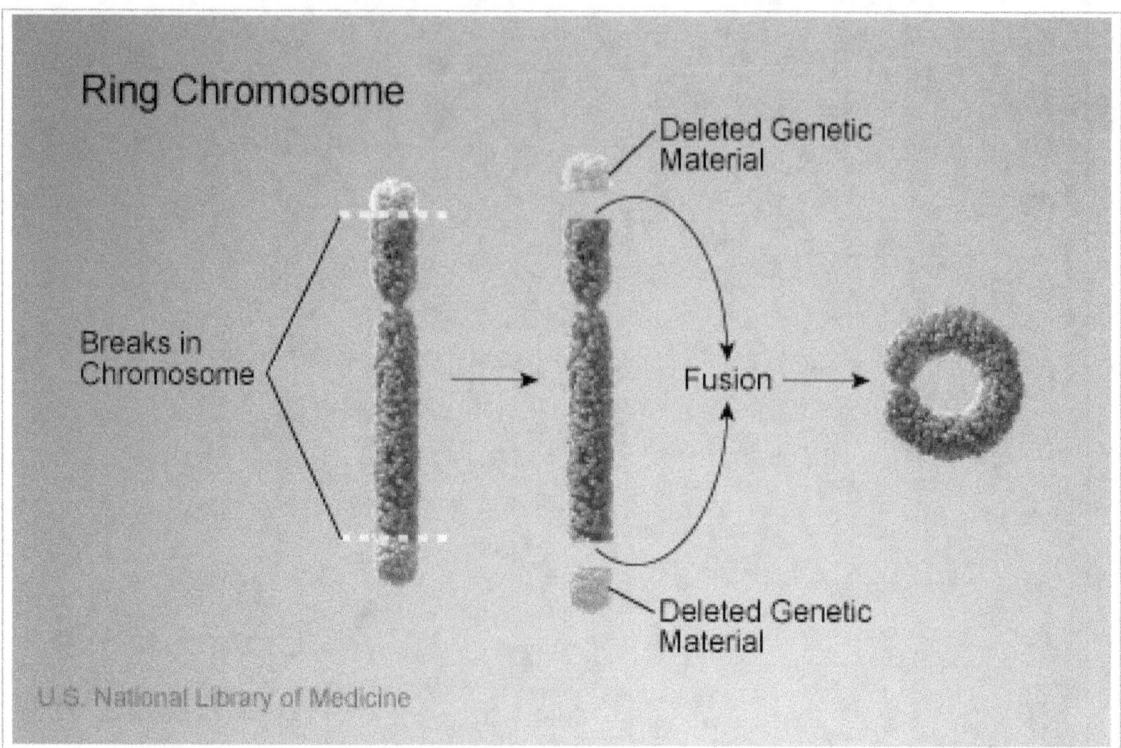

Ring chromosomes usually occur when a chromosome breaks in two places and the ends of the chromosome arms fuse together to form a circular structure.

Can changes in mitochondrial DNA affect health and development?

Mitochondria (illustration on page 7) are structures within cells that convert the energy from food into a form that cells can use. Although most DNA is packaged in chromosomes within the nucleus, mitochondria also have a small amount of their own DNA (known as mitochondrial DNA or mtDNA). In some cases, inherited changes in mitochondrial DNA can cause problems with growth, development, and function of the body's systems. These mutations disrupt the mitochondria's ability to generate energy efficiently for the cell.

Conditions caused by mutations in mitochondrial DNA often involve multiple organ systems. The effects of these conditions are most pronounced in organs and tissues that require a lot of energy (such as the heart, brain, and muscles). Although the health consequences of inherited mitochondrial DNA mutations vary widely, frequently observed features include muscle weakness and wasting, problems with movement, diabetes, kidney failure, heart disease, loss of intellectual functions (dementia), hearing loss, and abnormalities involving the eyes and vision.

Mitochondrial DNA is also prone to somatic mutations, which are not inherited. Somatic mutations occur in the DNA of certain cells during a person's lifetime and typically are not passed to future generations. Because mitochondrial DNA has a limited ability to repair itself when it is damaged, these mutations tend to build up over time. A buildup of somatic mutations in mitochondrial DNA has been associated with some forms of cancer and an increased risk of certain age-related disorders such as heart disease, Alzheimer disease, and Parkinson disease. Additionally, research suggests that the progressive accumulation of these mutations over a person's lifetime may play a role in the normal process of aging.

For more information about conditions caused by mitochondrial DNA mutations:

Genetics Home Reference provides background information about mitochondria and mitochondrial DNA (http://ghr.nlm.nih.gov/handbook/basics/mtdna) written in consumer-friendly language.

The Cleveland Clinic offers a basic introduction to mitochondrial disease (http://my.clevelandclinic.org/disorders/Mitochondrial_Disease/hic_Myths_and_Facts_About_Mitochondrial_Diseases.aspx). Additional information about mitochondrial disorders (http://www.cincinnatichildrens.org/service/m/mitochondrial-disorders/patients/) is available from Cincinnati Children's Hospital Medical Center.

An overview of mitochondrial disorders (http://www.ncbi.nlm.nih.gov/books/NBK1224/) is available from GeneReviews.

The Muscular Dystrophy Association offers an introduction to mitochondrial disorders as part of their fact sheet called Mitochondrial Myopathies (http://mda.org/disease/mitochondrial-myopathies).

The Neuromuscular Disease Center at Washington University provides an in-depth description of many mitochondrial conditions (http://neuromuscular.wustl.edu/mitosyn.html).

What are complex or multifactorial disorders?

Researchers are learning that nearly all conditions and diseases have a genetic component. Some disorders, such as sickle cell disease and cystic fibrosis, are caused by mutations in a single gene. The causes of many other disorders, however, are much more complex. Common medical problems such as heart disease, diabetes, and obesity do not have a single genetic cause—they are likely associated with the effects of multiple genes in combination with lifestyle and environmental factors. Conditions caused by many contributing factors are called complex or multifactorial disorders.

Although complex disorders often cluster in families, they do not have a clear-cut pattern of inheritance. This makes it difficult to determine a person's risk of inheriting or passing on these disorders. Complex disorders are also difficult to study and treat because the specific factors that cause most of these disorders have not yet been identified. Researchers continue to look for major contributing genes for many common complex disorders.

For more information about complex disorders:

A fact sheet about the inheritance of multifactorial disorders (http://www.genetics.edu.au/Publications-and-Resources/Genetics-Fact-Sheets/ Fact%20Sheet%2011) is available from the Centre for Genetics Education.

The Children's Hospital of Wisconsin provides basic information about multifactorial inheritance (http://www.chw.org/medical-care/genetics-and-genomics-program/ medical-genetics/multifactorial-inheritance/) and examples of multifactorial disorders.

Nature Education's Scitable offers a detailed description of complex and multifactorial diseases (http://www.nature.com/scitable/topicpage/complex-diseases-research-and-applications-748) and how researchers are studying them.

The National Human Genome Research Institute describes how researchers study complex disorders (http://www.genome.gov/10000865).

If you would like information about a specific complex disorder such as diabetes or obesity, MedlinePlus (http://www.nlm.nih.gov/medlineplus/) will lead you to fact sheets and other reliable medical information. In addition, the Centers for Disease Control and Prevention provides a detailed list of diseases and conditions (http://www.cdc.gov/DiseasesConditions/) that links to additional information.

What does it mean to have a genetic predisposition to a disease?

A genetic predisposition (sometimes also called genetic susceptibility) is an increased likelihood of developing a particular disease based on a person's genetic makeup. A genetic predisposition results from specific genetic variations that are often inherited from a parent. These genetic changes contribute to the development of a disease but do not directly cause it. Some people with a predisposing genetic variation will never get the disease while others will, even within the same family.

Genetic variations can have large or small effects on the likelihood of developing a particular disease. For example, certain mutations in the BRCA1 or BRCA2 genes greatly increase a person's risk of developing breast cancer and ovarian cancer. Variations in other genes, such as BARD1 and BRIP1, also increase breast cancer risk, but the contribution of these genetic changes to a person's overall risk appears to be much smaller.

Current research is focused on identifying genetic changes that have a small effect on disease risk but are common in the general population. Although each of these variations only slightly increases a person's risk, having changes in several different genes may combine to increase disease risk significantly. Changes in many genes, each with a small effect, may underlie susceptibility to many common diseases, including cancer, obesity, diabetes, heart disease, and mental illness.

In people with a genetic predisposition, the risk of disease can depend on multiple factors in addition to an identified genetic change. These include other genetic factors (sometimes called modifiers) as well as lifestyle and environmental factors. Diseases that are caused by a combination of factors are described as multifactorial (http://ghr.nlm.nih.gov/handbook/mutationsanddisorders/complexdisorders). Although a person's genetic makeup cannot be altered, some lifestyle and environmental modifications (such as having more frequent disease screenings and maintaining a healthy weight) may be able to reduce disease risk in people with a genetic predisposition.

For more information about genetic predisposition to disease:

The World Health Organization offers information about genetic predisposition to several common diseases (http://www.who.int/genomics/public/geneticdiseases/en/index3.html), including cancer, diabetes, cardiovascular disease, and asthma.

Genetic Alliance UK offers a fact sheet on genetic predisposition to common genetic diseases (http://www.geneticalliance.org.uk/education4.htm).

The Genetic Science Learning Center at the University of Utah provides more information about calculating the risk of genetic diseases and predicting disease based on family history (http://learn.genetics.utah.edu/content/history/geneticrisk/).

The Coriell Personalized Medicine Collaborative explains genetic and non-genetic risk factors (http://cpmc1.coriell.org/genetic-education/genetic-and-non-genetic-risk) for complex diseases.

More detailed information about the genetics of breast and ovarian cancer (http://www.cancer.gov/cancertopics/pdq/genetics/breast-and-ovarian/HealthProfessional/) is available from the National Cancer Institute.

What information about a genetic condition can statistics provide?

Statistical data can provide general information about how common a condition is, how many people have the condition, or how likely it is that a person will develop the condition. Statistics are not personalized, however—they offer estimates based on groups of people. By taking into account a person's family history, medical history, and other factors, a genetics professional can help interpret what statistics mean for a particular patient.

Some statistical terms are commonly used when describing genetic conditions and other disorders. These terms include:

Common statistical terms

Statistical term	Description	Examples
Incidence	The incidence of a gene mutation or a genetic disorder is the number of people who are born with the mutation or disorder in a specified group per year. Incidence is often written in the form "1 in [a number]" or as a total number of live births.	About 1 in 200,000 people in the United States are born with syndrome A each year. An estimated 15,000 infants with syndrome B were born last year worldwide.
Prevalence	The prevalence of a gene mutation or a genetic disorder is the total number of people in a specified group at a given time who have the mutation or disorder. This term includes both newly diagnosed and pre-existing cases in people of any age. Prevalence is often written in the form "1 in [a number]" or as a total number of people who have a condition.	Approximately 1 in 100,000 people in the United States have syndrome A at the present time. About 100,000 children worldwide currently have syndrome B.
Mortality	Mortality is the number of deaths from a particular disorder occurring in a specified group per year. Mortality is usually expressed as a total number of deaths.	An estimated 12,000 people worldwide died from syndrome C in 2002.
Lifetime risk	Lifetime risk is the average risk of developing a particular disorder at some point during a lifetime. Lifetime risk is often written as a percentage or as "1 in [a number]." It is important to remember that the risk per year or per decade is much lower than the lifetime risk. In addition, other factors may increase or decrease a person's risk as compared with the average.	Approximately 1 percent of people in the United States develop disorder D during their lifetimes. The lifetime risk of developing disorder D is 1 in 100.

For more information about understanding and interpreting statistics:

The New York Department of Health provides a basic explanation of statistical terms (http://www.health.ny.gov/diseases/chronic/basicstat.htm), including incidence, prevalence, morbidity, and mortality.

More detailed information about health statistics is available from Woloshin, Schwartz, and Welch's Know Your Chances: Understanding Health Statistics (http://www.ncbi.nlm.nih.gov/books/NBK115435/), which is available through the NCBI Bookshelf.

The National Cancer Institute offers additional tools for understanding cancer statistics (http://www.cancer.gov/statistics/understanding).

How are genetic conditions and genes named?

Naming genetic conditions

Genetic conditions are not named in one standard way (unlike genes, which are given an official name and symbol by a formal committee). Doctors who treat families with a particular disorder are often the first to propose a name for the condition. Expert working groups may later revise the name to improve its usefulness. Naming is important because it allows accurate and effective communication about particular conditions, which will ultimately help researchers find new approaches to treatment.

Disorder names are often derived from one or a combination of sources:

- The basic genetic or biochemical defect that causes the condition (for example, alpha-1 antitrypsin deficiency);

- One or more major signs or symptoms of the disorder (for example, hypermanganesemia with dystonia, polycythemia, and cirrhosis);

- The parts of the body affected by the condition (for example, craniofacial-deafness-hand syndrome);

- The name of a physician or researcher, often the first person to describe the disorder (for example, Marfan syndrome, which was named after Dr. Antoine Bernard-Jean Marfan);

- A geographic area (for example, familial Mediterranean fever, which occurs mainly in populations bordering the Mediterranean Sea); or

- The name of a patient or family with the condition (for example, amyotrophic lateral sclerosis, which is also called Lou Gehrig disease after the famous baseball player who had the condition).

Disorders named after a specific person or place are called eponyms. There is debate as to whether the possessive form (e.g., Alzheimer's disease) or the nonpossessive form (Alzheimer disease) of eponyms is preferred. As a rule, medical geneticists use the nonpossessive form, and this form may become the standard for doctors in all fields of medicine.

Naming genes

The HUGO Gene Nomenclature Committee (http://www.genenames.org/) (HGNC) designates an official name and symbol (an abbreviation of the name) for each known human gene. Some official gene names include additional information in parentheses, such as related genetic conditions, subtypes of a condition, or inheritance pattern. The HGNC is a non-profit organization funded by the U.K. Medical Research Council and the U.S. National Institutes of Health. The Committee has named more than 13,000 of the estimated 20,000 to 25,000 genes in the human genome.

During the research process, genes often acquire several alternate names and symbols. Different researchers investigating the same gene may each give the gene a different name, which can cause confusion. The HGNC assigns a unique name and symbol to each human gene, which allows effective organization of genes in large databanks, aiding the advancement of research. For specific information about how genes are named, refer to the HGNC's Guidelines for Human Gene Nomenclature (http://www.genenames.org/guidelines.html).

Chapter 4

Inheriting Genetic Conditions

Table of Contents

What does it mean if a disorder seems to run in my family?

A particular disorder might be described as "running in a family" if more than one person in the family has the condition. Some disorders that affect multiple family members are caused by gene mutations, which can be inherited (passed down from parent to child). Other conditions that appear to run in families are not caused by mutations in single genes. Instead, environmental factors such as dietary habits or a combination of genetic and environmental factors are responsible for these disorders.

It is not always easy to determine whether a condition in a family is inherited. A genetics professional can use a person's family history (a record of health information about a person's immediate and extended family) to help determine whether a disorder has a genetic component. He or she will ask about the health of people from several generations of the family, usually first-, second-, and third-degree relatives.

Degrees of relationship

Degrees of relationship	Examples
First-degree relatives	Parents, children, brothers, and sisters
Second-degree relatives	Grandparents, aunts and uncles, nieces and nephews, and grandchildren
Third-degree relatives	First cousin

Condition affecting members of a family

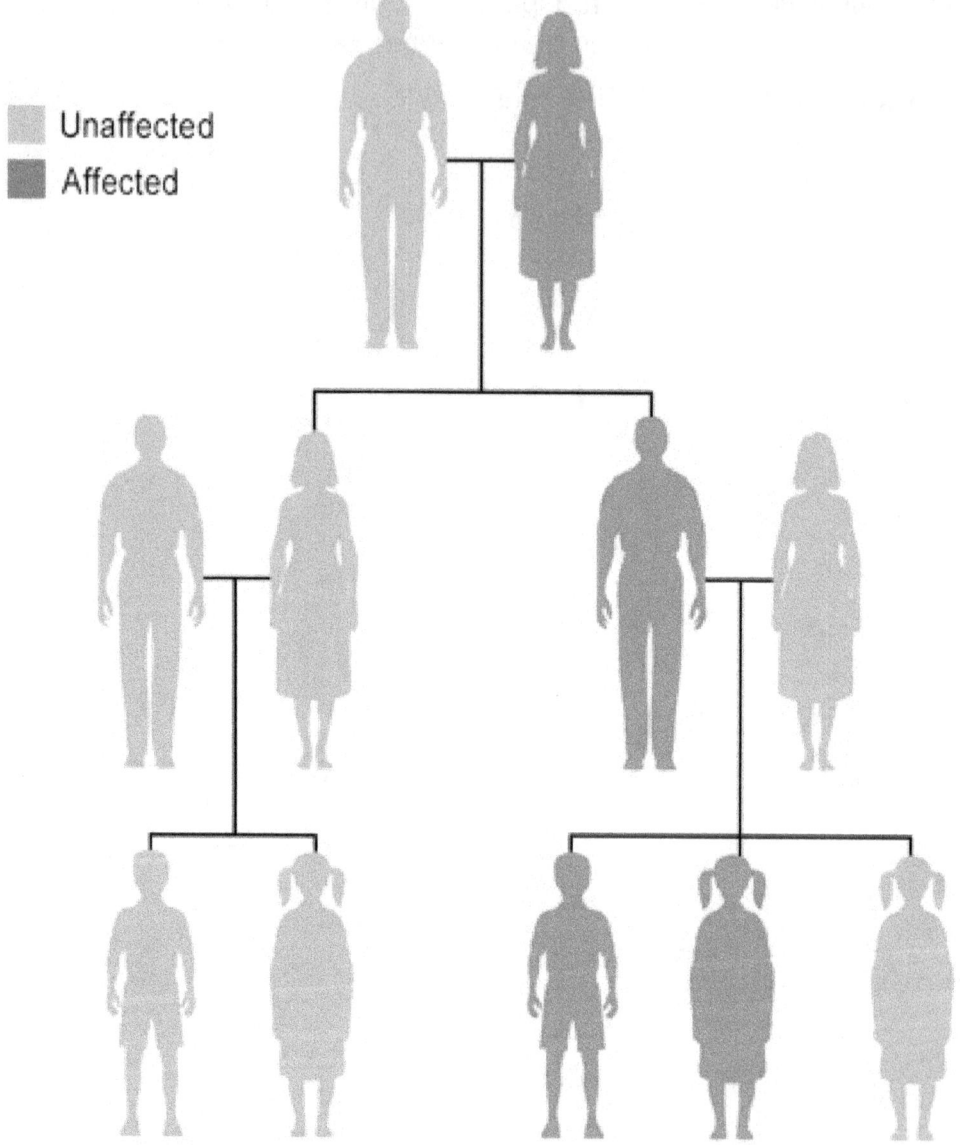

Unaffected
Affected

Some disorders are seen in more than one generation of a family.

For general information about disorders that run in families:

Genetics Home Reference provides consumer-friendly summaries of genetic conditions (http://ghr.nlm.nih.gov/BrowseConditions). Each summary includes a brief description of the condition, an explanation of its genetic cause, and information about the condition's frequency and pattern of inheritance.

The Coriell Personalized Medicine Collaborative provides a brief introduction to heritable diseases in the article Heredity: It Runs in the Family (http://cpmc1.coriell.org/genetic-education/it-runs-in-the-family).

The Genetic Science Learning Center at the University of Utah offers interactive tools about disorders that run in families (http://learn.genetics.utah.edu/content/history).

The National Human Genome Research Institute offers a brief fact sheet called Frequently Asked Questions About Genetic Disorders (http://www.genome.gov/19016930).

The Centre for Genetics Education provides an overview of genetic conditions (http://www.genetics.edu.au/Publications-and-Resources/Genetics-Fact-Sheets/FactSheet2).

Why is it important to know my family medical history?

A family medical history is a record of health information about a person and his or her close relatives. A complete record includes information from three generations of relatives, including children, brothers and sisters, parents, aunts and uncles, nieces and nephews, grandparents, and cousins.

Families have many factors in common, including their genes, environment, and lifestyle. Together, these factors can give clues to medical conditions that may run in a family. By noticing patterns of disorders among relatives, healthcare professionals can determine whether an individual, other family members, or future generations may be at an increased risk of developing a particular condition.

A family medical history can identify people with a higher-than-usual chance of having common disorders, such as heart disease, high blood pressure, stroke, certain cancers, and diabetes. These complex disorders are influenced by a combination of genetic factors, environmental conditions, and lifestyle choices. A family history also can provide information about the risk of rarer conditions caused by mutations in a single gene, such as cystic fibrosis and sickle cell anemia.

While a family medical history provides information about the risk of specific health concerns, having relatives with a medical condition does not mean that an individual will definitely develop that condition. On the other hand, a person with no family history of a disorder may still be at risk of developing that disorder.

Knowing one's family medical history allows a person to take steps to reduce his or her risk. For people at an increased risk of certain cancers, healthcare professionals may recommend more frequent screening (such as mammography or colonoscopy) starting at an earlier age. Healthcare providers may also encourage regular checkups or testing for people with a medical condition that runs in their family. Additionally, lifestyle changes such as adopting a healthier diet, getting regular exercise, and quitting smoking help many people lower their chances of developing heart disease and other common illnesses.

The easiest way to get information about family medical history is to talk to relatives about their health. Have they had any medical problems, and when did they occur? A family gathering could be a good time to discuss these issues. Additionally, obtaining medical records and other documents (such as obituaries and death certificates) can help complete a family medical history. It is important to keep this information up-to-date and to share it with a healthcare professional regularly.

For more information about family medical history:

NIHSeniorHealth, a service of the National Institutes of Health, provides information and tools (http://nihseniorhealth.gov/creatingafamilyhealthhistory/

whycreateafamilyhealthhistory/01.html) for documenting family health history. Additional information about family history (http://www.nlm.nih.gov/medlineplus/familyhistory.html) is available from MedlinePlus.

Educational resources related to family health history (http://geneed.nlm.nih.gov/topic_subtopic.php?tid=5&sid=13) are available from GeneEd.

The Centers for Disease Control and Prevention's (CDC) of Public Health Genomics provides information about the importance of family medical history (http://www.cdc.gov/genomics/famhistory/famhist.htm). This resource also includes links to publications, reports, and tools for recording family health information.

The Office of the Surgeon General offers a tool called My Family Health Portrait (http://familyhistory.hhs.gov/) that allows you to enter, print, and update your family health history.

Information about collecting and recording a family medical history (http://nsgc.org/p/cm/ld/fid=52) is also available from the National Society of Genetic Counselors.

The American Medical Association provides family history tools (http://www.ama-assn.org/ama/pub/physician-resources/medical-science/genetics-molecular-medicine/family-history.page), including questionnaires and forms for collecting medical information.

The National Genetics and Genomics Education Centre of the National Health Service (UK) describes how healthcare providers collect information about a person's family health history (http://www.geneticseducation.nhs.uk/for-practitioners-62/identifying-patients/taking-and-recording-a-family-history).

Links to additional resources (http://www.kumc.edu/gec/pedigree.html) are available from the University of Kansas Medical Center. The Genetic Alliance also offers a list of links to family history resources (http://www.geneticalliance.org/fhh).

What are the different ways in which a genetic condition can be inherited?

Some genetic conditions are caused by mutations in a single gene. These conditions are usually inherited in one of several straightforward patterns, depending on the gene involved:

Patterns of inheritance

Inheritance pattern	Description	Examples
Autosomal dominant	One mutated copy of the gene in each cell is sufficient for a person to be affected by an autosomal dominant disorder. Each affected person usually has one affected parent (illustration on page 89). Autosomal dominant disorders tend to occur in every generation of an affected family.	Huntington disease, neurofibromatosis type 1
Autosomal recessive	Two mutated copies of the gene are present in each cell when a person has an autosomal recessive disorder. An affected person usually has unaffected parents who each carry a single copy of the mutated gene (and are referred to as carriers) (illustration on page 90). Autosomal recessive disorders are typically not seen in every generation of an affected family.	cystic fibrosis, sickle cell anemia
X-linked dominant	X-linked dominant disorders are caused by mutations in genes on the X chromosome. Females are more frequently affected than males, and the chance of passing on an X-linked dominant disorder differs between men (illustration on page 91) and women (illustration on page 92). Families with an X-linked dominant disorder often have both affected males and affected females in each generation. A characteristic of X-linked inheritance is that fathers cannot pass X-linked traits to their sons (no male-to-male transmission).	fragile X syndrome
X-linked recessive		hemophilia, Fabry disease

Inheritance pattern	Description	Examples
	X-linked recessive disorders are also caused by mutations in genes on the X chromosome. Males are more frequently affected than females, and the chance of passing on the disorder differs between men (illustration on page 93) and women (illustration on page 94). Families with an X-linked recessive disorder often have affected males, but rarely affected females, in each generation. A characteristic of X-linked inheritance is that fathers cannot pass X-linked traits to their sons (no male-to-male transmission).	
Codominant	In codominant inheritance, two different versions (alleles) of a gene can be expressed, and each version makes a slightly different protein (illustration on page 95). Both alleles influence the genetic trait or determine the characteristics of the genetic condition.	ABO blood group, alpha-1 antitrypsin deficiency
Mitochondrial	This type of inheritance, also known as maternal inheritance, applies to genes in mitochondrial DNA. Mitochondria, which are structures in each cell that convert molecules into energy, each contain a small amount of DNA. Because only egg cells contribute mitochondria to the developing embryo, only females can pass on mitochondrial mutations to their children (illustration on page 96). Disorders resulting from mutations in mitochondrial DNA can appear in every generation of a family and can affect both males and females, but fathers do not pass these disorders to their children.	Leber hereditary optic neuropathy (LHON)

Many other disorders are caused by a combination of the effects of multiple genes or by interactions between genes and the environment. Such disorders are more difficult to analyze because their genetic causes are often unclear, and they do not follow the patterns of inheritance described above. Examples of conditions caused by multiple genes or gene/environment interactions include heart disease, diabetes, schizophrenia, and certain types of cancer. For more information, please see What are complex or multifactorial disorders? (http://ghr.nlm.nih.gov/handbook/mutationsanddisorders/complexdisorders).

Disorders caused by changes in the number or structure of chromosomes do not follow the straightforward patterns of inheritance listed above. To read about how chromosomal conditions occur, please see Are chromosomal disorders inherited? (http://ghr.nlm.nih.gov/handbook/inheritance/chromosomalinheritance).

Other genetic factors can also influence how a disorder is inherited: What are genomic imprinting and uniparental disomy? (http://ghr.nlm.nih.gov/handbook/inheritance/updimprinting)

For more information about inheritance patterns:

The Genetics and Public Policy Center provides an introduction to genetic inheritance patterns (http://www.dnapolicy.org/science.gh.php).

Resources related to heredity/inheritance patterns (http://geneed.nlm.nih.gov/topic_subtopic.php?tid=5) and Mendelian inheritance (http://geneed.nlm.nih.gov/topic_subtopic.php?tid=5&sid=6) are available from GeneEd.

The Centre for Genetics Education provides information about each of the inheritance patterns outlined above:

- Autosomal dominant inheritance (http://www.genetics.edu.au/Publications-and-Resources/Genetics-Fact-Sheets/Fact%20Sheet%209)

- Autosomal recessive inheritance (http://www.genetics.edu.au/Publications-and-Resources/Genetics-Fact-Sheets/FactSheet8)

- X-linked dominant inheritance (http://www.genetics.edu.au/Publications-and-Resources/Genetics-Fact-Sheets/FactSheet10A)

- X-linked recessive inheritance (http://www.genetics.edu.au/Publications-and-Resources/Genetics-Fact-Sheets/Fact%20Sheet%2010)

- Mitochondrial inheritance (http://www.genetics.edu.au/Publications-and-Resources/Genetics-Fact-Sheets/Fact%20Sheet%2012)

EuroGentest also offers explanations of Mendelian inheritance patterns:

- Autosomal dominant inheritance (http://www.eurogentest.org/index.php?id=614)

- Autosomal recessive inheritance (http://www.eurogentest.org/index.php?id=619)

- X-linked inheritance (http://www.eurogentest.org/index.php?id=623)

The Education Portal offers detailed information about different inheritance patterns:

- Overview of inheritance patterns (http://education-portal.com/academy/lesson/pedigree-analysis-in-human-genetics-inheritance-patterns.html#lesson)

- Sex-linked and sex-limited traits (http://education-portal.com/academy/lesson/exceptions-to-independent-assortment-sex-linked-and-sex-limited-traits.html#lesson)

Additional information about inheritance patterns is available from The Merck Manual (http://www.merckmanuals.com/professional/special_subjects/general_principles_of_medical_genetics/single-gene_defects.html).

The National Genetics and Genomics Education Centre of the National Health Service (UK) provides explanations of various forms of genetic inheritance (http://www.geneticseducation.nhs.uk/for-practitioners-62/communicating-information/explaining-inheritance).

Illustrations

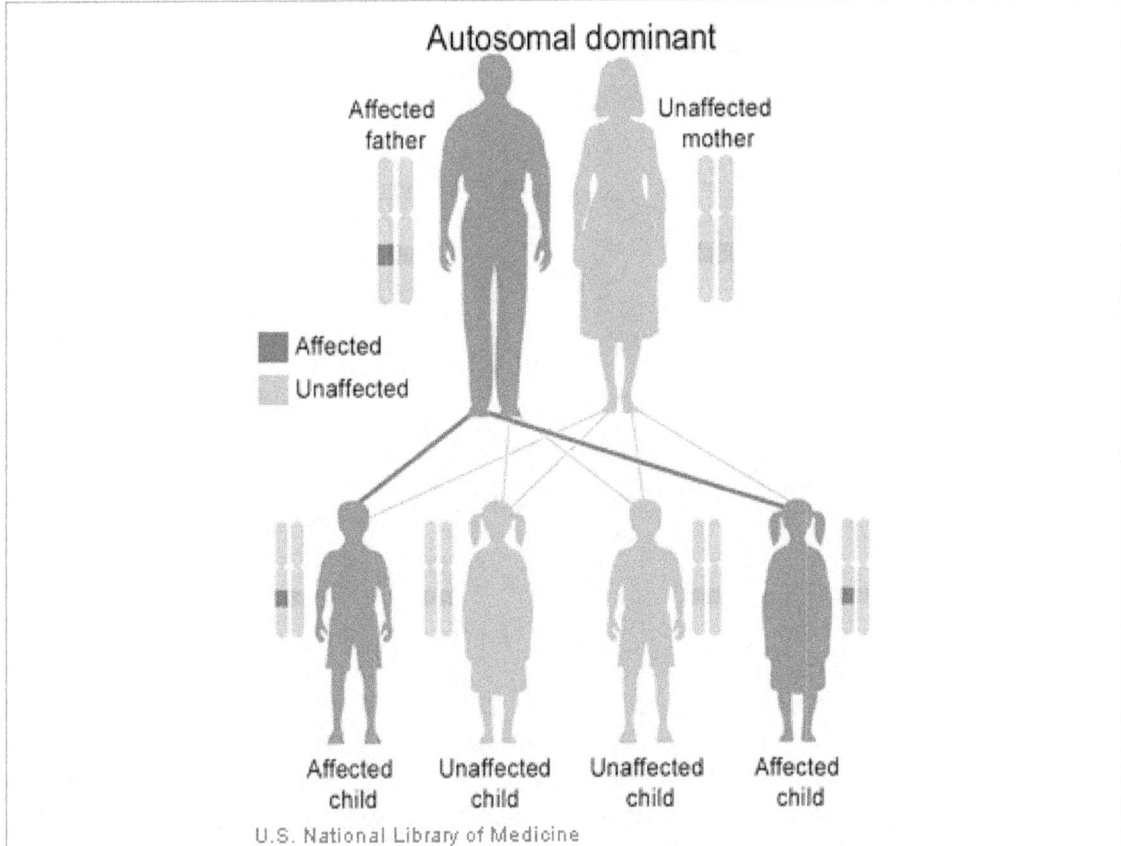

In this example, a man with an autosomal dominant disorder has two affected children and two unaffected children.

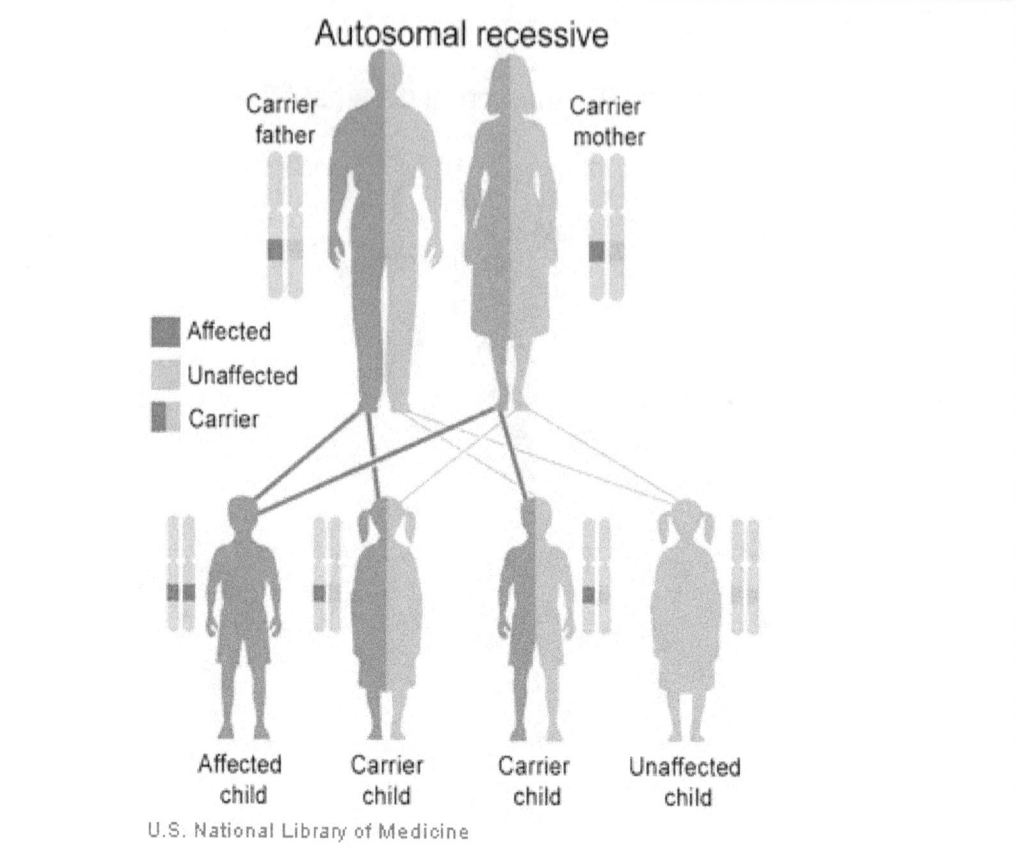

Autosomal recessive

In this example, two unaffected parents each carry one copy of a gene mutation for an autosomal recessive disorder. They have one affected child and three unaffected children, two of which carry one copy of the gene mutation.

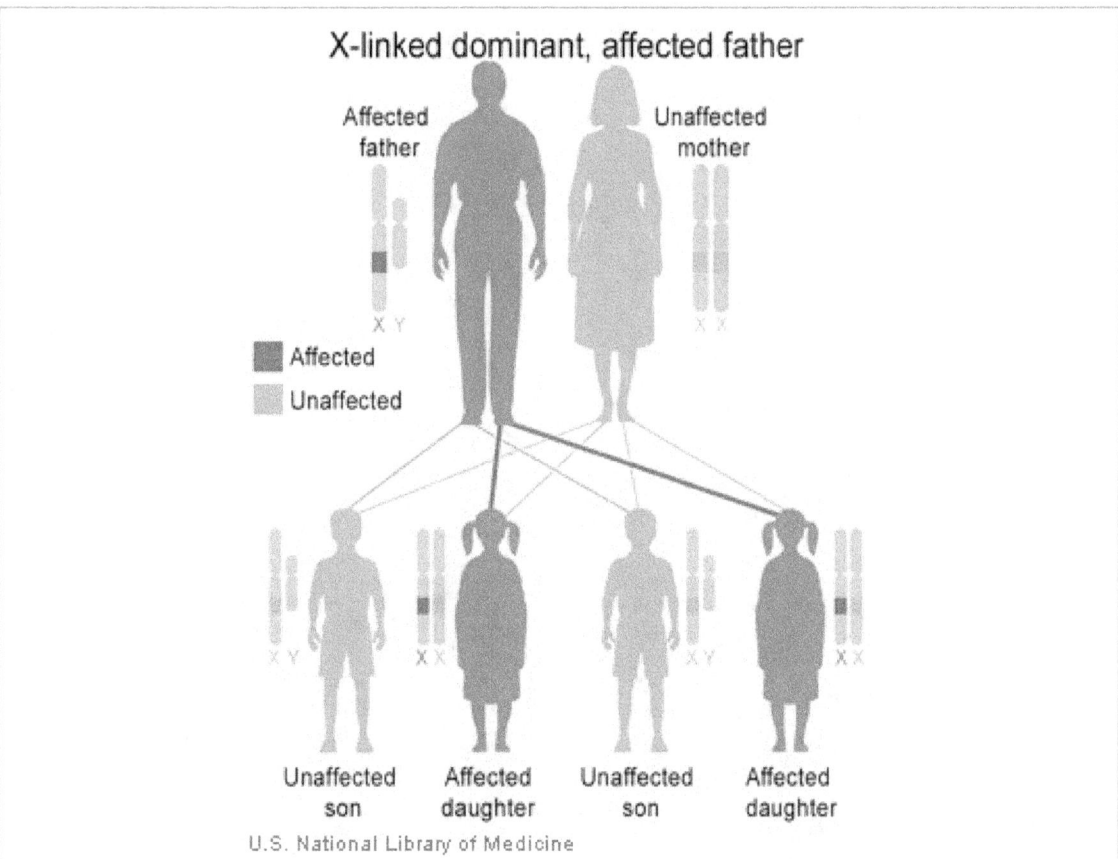

X-linked dominant, affected father

Affected father

Unaffected mother

Affected
Unaffected

Unaffected son

Affected daughter

Unaffected son

Affected daughter

U.S. National Library of Medicine

In this example, a man with an X-linked dominant condition has two affected daughters and two unaffected sons.

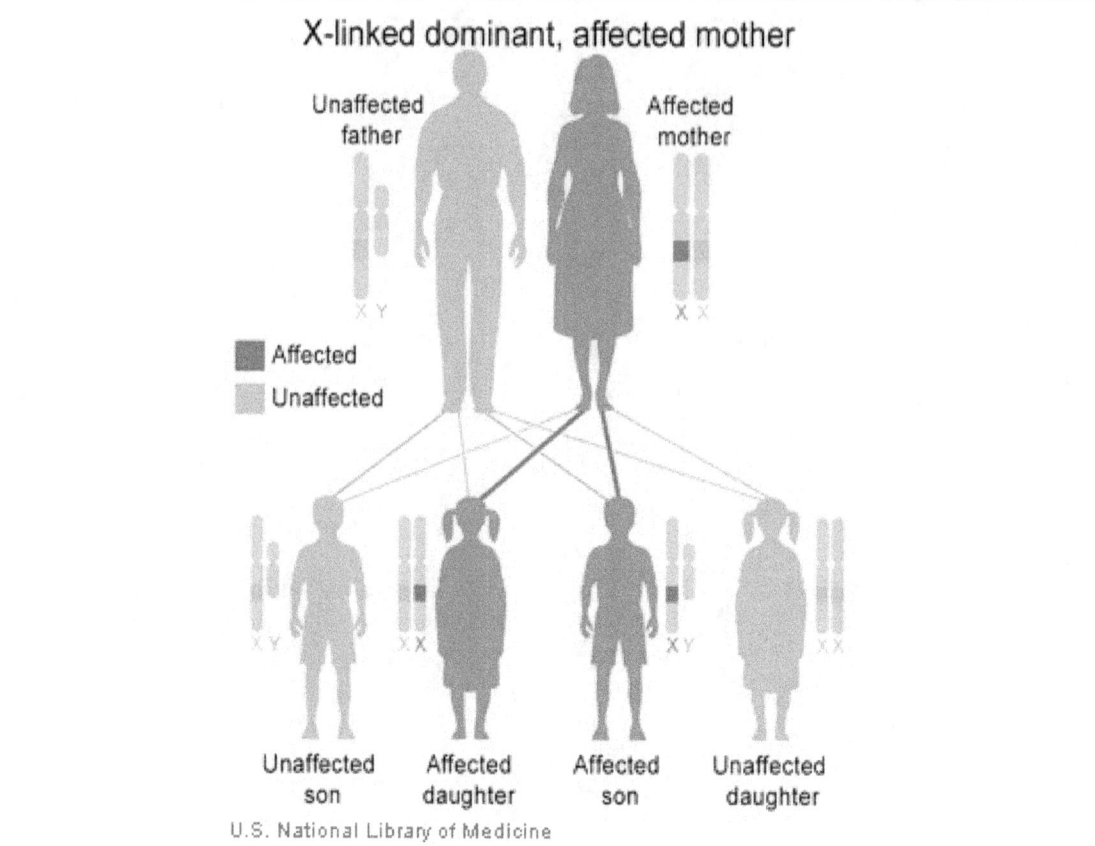

In this example, a woman with an X-linked dominant condition has an affected daughter, an affected son, an unaffected daughter, and an unaffected son.

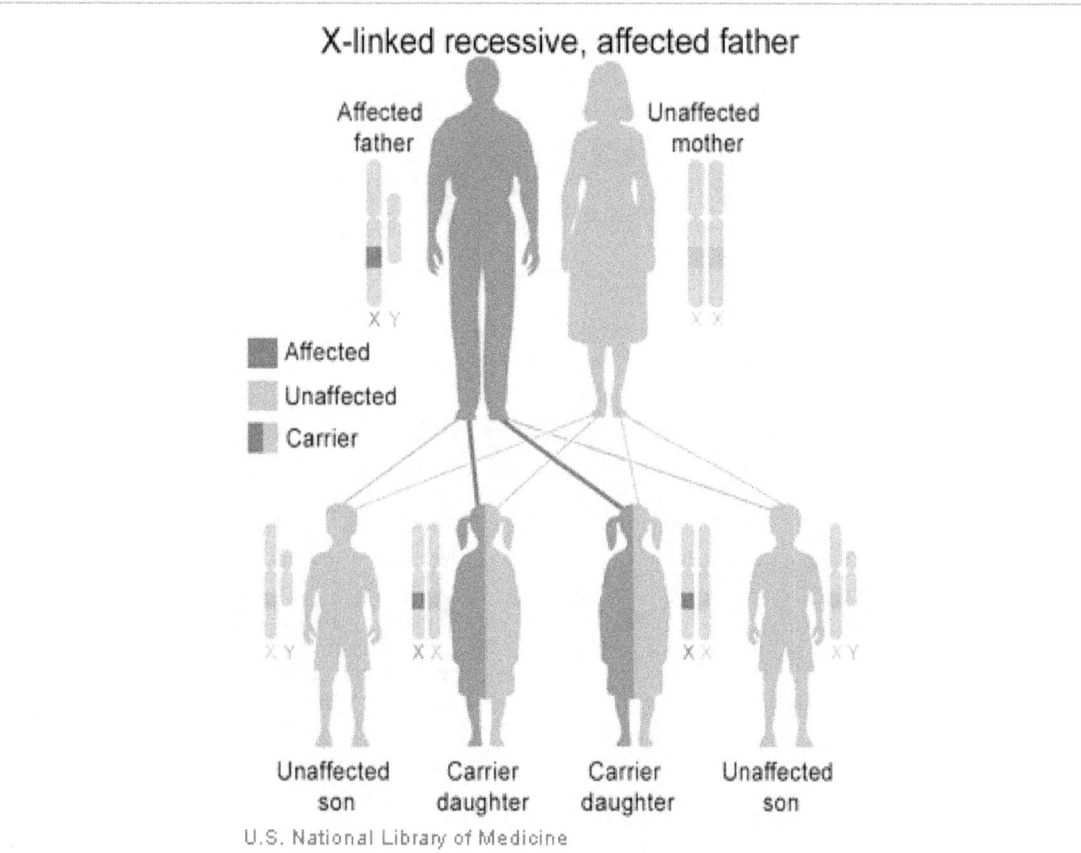

X-linked recessive, affected father

Affected father — X Y

Unaffected mother — X X

■ Affected
■ Unaffected
■ Carrier

Unaffected son — X Y

Carrier daughter — X X

Carrier daughter — X X

Unaffected son — X Y

U.S. National Library of Medicine

In this example, a man with an X-linked recessive condition has two unaffected daughters who each carry one copy of the gene mutation, and two unaffected sons who do not have the mutation.

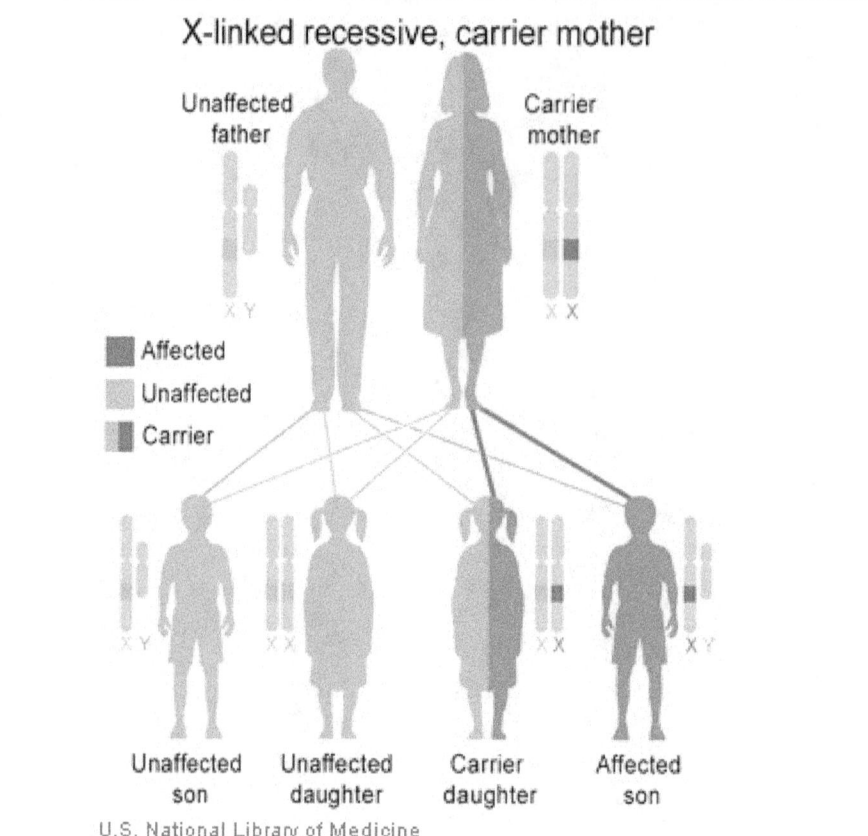

X-linked recessive, carrier mother

Unaffected father

Carrier mother

X Y   X X

■ Affected
□ Unaffected
■ Carrier

X Y   X X   X X   X Y

Unaffected son

Unaffected daughter

Carrier daughter

Affected son

In this example, an unaffected woman carries one copy of a gene mutation for an X-linked recessive disorder. She has an affected son, an unaffected daughter who carries one copy of the mutation, and two unaffected children who do not have the mutation.

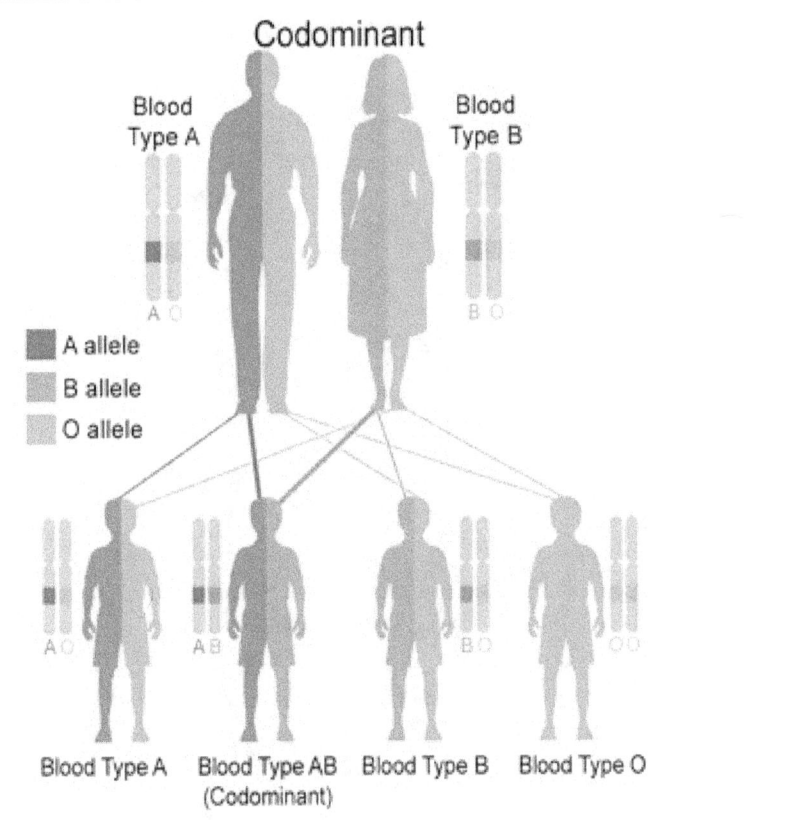

Codominant

Blood Type A

Blood Type B

- ■ A allele
- ■ B allele
- ■ O allele

A O   B O

A O   A B   B O   O O

Blood Type A   Blood Type AB   Blood Type B   Blood Type O
(Codominant)

U.S. National Library of Medicine

The ABO blood group is a major system for classifying blood types in humans. Blood type AB is inherited in a codominant pattern. In this example, a father with blood type A and a mother with blood type B have four children, each with a different blood type: A, AB, B, and O.

Mitochondrial

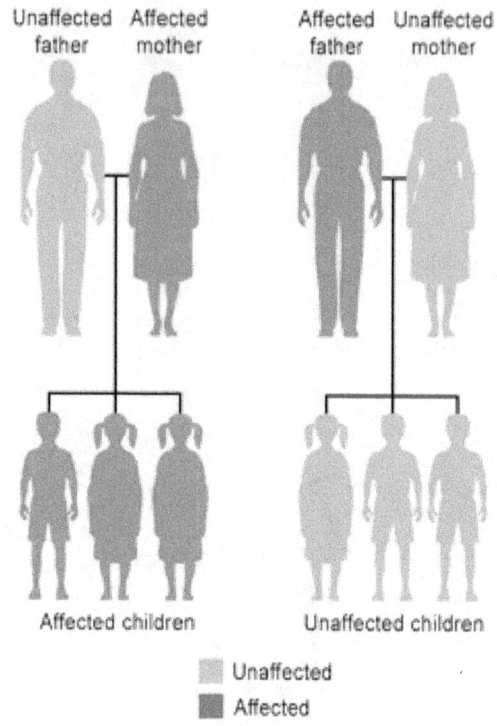

Unaffected father   Affected mother

Affected father   Unaffected mother

Affected children

Unaffected children

Unaffected
Affected

In one family, a woman with a disorder caused by a mutation in mitochondrial DNA and her unaffected husband have only affected children. In another family, a man with a condition resulting from a mutation in mitochondrial DNA and his unaffected wife have no affected children.

If a genetic disorder runs in my family, what are the chances that my children will have the condition?

When a genetic disorder is diagnosed in a family, family members often want to know the likelihood that they or their children will develop the condition. This can be difficult to predict in some cases because many factors influence a person's chances of developing a genetic condition. One important factor is how the condition is inherited. For example:

- Autosomal dominant inheritance: A person affected by an autosomal dominant disorder has a 50 percent chance of passing the mutated gene to each child. The chance that a child will not inherit the mutated gene is also 50 percent (illustration on page 89).

- Autosomal recessive inheritance: Two unaffected people who each carry one copy of the mutated gene for an autosomal recessive disorder (carriers) have a 25 percent chance with each pregnancy of having a child affected by the disorder. The chance with each pregnancy of having an unaffected child who is a carrier of the disorder is 50 percent, and the chance that a child will not have the disorder and will not be a carrier is 25 percent (illustration on page 90).

- X-linked dominant inheritance: The chance of passing on an X-linked dominant condition differs between men and women because men have one X chromosome and one Y chromosome, while women have two X chromosomes. A man passes on his Y chromosome to all of his sons and his X chromosome to all of his daughters. Therefore, the sons of a man with an X-linked dominant disorder will not be affected, but all of his daughters will inherit the condition (illustration on page 91). A woman passes on one or the other of her X chromosomes to each child. Therefore, a woman with an X-linked dominant disorder has a 50 percent chance of having an affected daughter or son with each pregnancy (illustration on page 92).

- X-linked recessive inheritance: Because of the difference in sex chromosomes, the probability of passing on an X-linked recessive disorder also differs between men and women. The sons of a man with an X-linked recessive disorder will not be affected, and his daughters will carry one copy of the mutated gene (illustration on page 93). With each pregnancy, a woman who carries an X-linked recessive disorder has a 50 percent chance of having sons who are affected and a 50 percent chance of having daughters who carry one copy of the mutated gene (illustration on page 94).

- Codominant inheritance: In codominant inheritance, each parent contributes a different version of a particular gene, and both versions influence the resulting genetic trait. The chance of developing a genetic condition with codominant inheritance, and the characteristic features of that condition, depend on which versions of the gene are passed from parents to their child (illustration on page 95).

- Mitochondrial inheritance: Mitochondria, which are the energy-producing centers inside cells, each contain a small amount of DNA. Disorders with mitochondrial inheritance result from mutations in mitochondrial DNA. Although these disorders can affect both males and females, only females can pass mutations in mitochondrial DNA to their children. A woman with a disorder caused by changes in mitochondrial DNA will pass the mutation to all of her daughters and sons, but the children of a man with such a disorder will not inherit the mutation (illustration on page 96).

It is important to note that the chance of passing on a genetic condition applies equally to each pregnancy. For example, if a couple has a child with an autosomal recessive disorder, the chance of having another child with the disorder is still 25 percent (or 1 in 4). Having one child with a disorder does not "protect" future children from inheriting the condition. Conversely, having a child without the condition does not mean that future children will definitely be affected.

Although the chances of inheriting a genetic condition appear straightforward, factors such as a person's family history and the results of genetic testing can sometimes modify those chances. In addition, some people with a disease-causing mutation never develop any health problems or may experience only mild symptoms of the disorder. If a disease that runs in a family does not have a clear-cut inheritance pattern, predicting the likelihood that a person will develop the condition can be particularly difficult.

Estimating the chance of developing or passing on a genetic disorder can be complex. Genetics professionals can help people understand these chances and help them make informed decisions about their health.

For more information about passing on a genetic disorder in a family:

The National Library of Medicine MedlinePlus web site offers information about the chance of developing a genetic disorder on the basis of its inheritance pattern. Scroll down to the section "Statistical Chances of Inheriting a Trait" for each of the following inheritance patterns:

- Autosomal dominant (http://www.nlm.nih.gov/medlineplus/ency/article/002049.htm)

- Autosomal recessive (http://www.nlm.nih.gov/medlineplus/ency/article/002052.htm)

- X-linked dominant (http://www.nlm.nih.gov/medlineplus/ency/article/002050.htm)

- X-linked recessive (http://www.nlm.nih.gov/medlineplus/ency/article/002051.htm)

The Centre for Genetics Education (Australia) provides an explanation of mitochondrial inheritance (http://www.genetics.edu.au/Publications-and-Resources/Genetics-Fact-Sheets/Fact%20Sheet%2012).

The Muscular Dystrophy Association explains patterns and probabilities (http://mda.org/publications/facts-about-genetics-and-NMDs/genetic-disorders-inherited) of inheritance.

What are reduced penetrance and variable expressivity?

Reduced penetrance and variable expressivity are factors that influence the effects of particular genetic changes. These factors usually affect disorders that have an autosomal dominant pattern of inheritance, although they are occasionally seen in disorders with an autosomal recessive inheritance pattern.

Reduced penetrance

Penetrance refers to the proportion of people with a particular genetic change (such as a mutation in a specific gene) who exhibit signs and symptoms of a genetic disorder. If some people with the mutation do not develop features of the disorder, the condition is said to have reduced (or incomplete) penetrance. Reduced penetrance often occurs with familial cancer syndromes. For example, many people with a mutation in the BRCA1 or BRCA2 gene will develop cancer during their lifetime, but some people will not. Doctors cannot predict which people with these mutations will develop cancer or when the tumors will develop.

Reduced penetrance probably results from a combination of genetic, environmental, and lifestyle factors, many of which are unknown. This phenomenon can make it challenging for genetics professionals to interpret a person's family medical history and predict the risk of passing a genetic condition to future generations.

Variable expressivity

Although some genetic disorders exhibit little variation, most have signs and symptoms that differ among affected individuals. Variable expressivity refers to the range of signs and symptoms that can occur in different people with the same genetic condition. For example, the features of Marfan syndrome vary widely— some people have only mild symptoms (such as being tall and thin with long, slender fingers), while others also experience life-threatening complications involving the heart and blood vessels. Although the features are highly variable, most people with this disorder have a mutation in the same gene (FBN1).

As with reduced penetrance, variable expressivity is probably caused by a combination of genetic, environmental, and lifestyle factors, most of which have not been identified. If a genetic condition has highly variable signs and symptoms, it may be challenging to diagnose.

For more information about reduced penetrance and variable expressivity:

The PHG Foundation offers an interactive tutorial on penetrance (http://www.phgfoundation.org/tutorials/penetrance/index.html) that explains the differences between reduced penetrance and variable expressivity.

The Education Portal provides a detailed discussion of penetrance and expressivity (http://education-portal.com/academy/lesson/genetic-disorders-penetrance-phenotypic-variability.html#lesson).

Additional information about penetrance and expressivity (http://www.merckmanuals.com/home/fundamentals/genetics/inheritance_of_single-gene_disorders.html) is available from the Merck Manual Home Health Handbook for Patients & Caregivers.

A more in-depth explanation of these concepts is available from the textbook Human Molecular Genetics 2 in chapter 3.2, Complications to the Basic Pedigree Patterns (http://www.ncbi.nlm.nih.gov/books/NBK7573/?redirect-on-error=__HOME__#A286).

What do geneticists mean by anticipation?

The signs and symptoms of some genetic conditions tend to become more severe and appear at an earlier age as the disorder is passed from one generation to the next. This phenomenon is called anticipation. Anticipation is most often seen with certain genetic disorders of the nervous system, such as Huntington disease, myotonic dystrophy, and fragile X syndrome.

Anticipation typically occurs with disorders that are caused by an unusual type of mutation called a trinucleotide repeat expansion. A trinucleotide repeat is a sequence of three DNA building blocks (nucleotides) that is repeated a number of times in a row. DNA segments with an abnormal number of these repeats are unstable and prone to errors during cell division. The number of repeats can change as the gene is passed from parent to child. If the number of repeats increases, it is known as a trinucleotide repeat expansion. In some cases, the trinucleotide repeat may expand until the gene stops functioning normally. This expansion causes the features of some disorders to become more severe with each successive generation.

Most genetic disorders have signs and symptoms that differ among affected individuals, including affected people in the same family. Not all of these differences can be explained by anticipation. A combination of genetic, environmental, and lifestyle factors is probably responsible for the variability, although many of these factors have not been identified. Researchers study multiple generations of affected family members and consider the genetic cause of a disorder before determining that it shows anticipation.

For more information about anticipation:

The Merck Manual for Healthcare Professionals provides a brief explanation of anticipation as part of its chapter on nontraditional inheritance (http://www.merckmanuals.com/professional/special_subjects/general_principles_of_medical_genetics/unusual_aspects_of_inheritance.html?qt=&sc=&alt=#v1123535).

The Myotonic Dystrophy Foundation describes anticipation in the context of myotonic dystrophy (http://www.myotonic.org/what-dm/disease-mechanism). (Click on the tab that says "Anticipation.")

Additional information about anticipation is available from the textbook Human Molecular Genetics 2 in chapter 3.2, Complications to the Basic Pedigree Patterns (http://www.ncbi.nlm.nih.gov/books/NBK7573/?redirect-on-error=__HOME__#A286).

What are genomic imprinting and uniparental disomy?

Genomic imprinting and uniparental disomy are factors that influence how some genetic conditions are inherited.

Genomic imprinting

People inherit two copies of their genes—one from their mother and one from their father. Usually both copies of each gene are active, or "turned on," in cells. In some cases, however, only one of the two copies is normally turned on. Which copy is active depends on the parent of origin: some genes are normally active only when they are inherited from a person's father; others are active only when inherited from a person's mother. This phenomenon is known as genomic imprinting.

In genes that undergo genomic imprinting, the parent of origin is often marked, or "stamped," on the gene during the formation of egg and sperm cells. This stamping process, called methylation, is a chemical reaction that attaches small molecules called methyl groups to certain segments of DNA. These molecules identify which copy of a gene was inherited from the mother and which was inherited from the father. The addition and removal of methyl groups can be used to control the activity of genes.

Only a small percentage of all human genes undergo genomic imprinting. Researchers are not yet certain why some genes are imprinted and others are not. They do know that imprinted genes tend to cluster together in the same regions of chromosomes. Two major clusters of imprinted genes have been identified in humans, one on the short (p) arm of chromosome 11 (at position 11p15) and another on the long (q) arm of chromosome 15 (in the region 15q11 to 15q13).

Uniparental disomy

Uniparental disomy (UPD) occurs when a person receives two copies of a chromosome, or part of a chromosome, from one parent and no copies from the other parent. UPD can occur as a random event during the formation of egg or sperm cells or may happen in early fetal development.

In many cases, UPD likely has no effect on health or development. Because most genes are not imprinted, it doesn't matter if a person inherits both copies from one parent instead of one copy from each parent. In some cases, however, it does make a difference whether a gene is inherited from a person's mother or father. A person with UPD may lack any active copies of essential genes that undergo genomic imprinting. This loss of gene function can lead to delayed development, mental retardation, or other medical problems.

Several genetic disorders can result from UPD or a disruption of normal genomic imprinting. The most well-known conditions include Prader-Willi syndrome, which is characterized by uncontrolled eating and obesity, and Angelman syndrome, which causes mental retardation and impaired speech. Both of these disorders can be caused by UPD or other errors in imprinting involving genes on the long arm of chromosome 15. Other conditions, such as Beckwith-Wiedemann syndrome (a disorder characterized by accelerated growth and an increased risk of cancerous tumors), are associated with abnormalities of imprinted genes on the short arm of chromosome 11.

For more information about genomic imprinting and UPD:

The University of Utah offers a basic overview of genomic imprinting (http://learn.genetics.utah.edu/content/epigenetics/imprinting/).

Additional information about genomic imprinting (http://www.genetics.edu.au/Publications-and-Resources/Genetics-Fact-Sheets/FactSheet15) is available from the Centre for Genetics Education.

Nature Education's Scitable provides additional information in the article "Genomic Imprinting and Patterns of Disease Inheritance." (http://www.nature.com/scitable/topicpage/genomic-imprinting-and-patterns-of-disease-inheritance-899)

An animated tutorial from the University of Miami illustrates how uniparental disomy occurs (http://hihg.med.miami.edu/code/http/modules/education/Design/animate/uniDisomy.htm).

Are chromosomal disorders inherited?

Although it is possible to inherit some types of chromosomal abnormalities, most chromosomal disorders (such as Down syndrome and Turner syndrome) are not passed from one generation to the next.

Some chromosomal conditions are caused by changes in the number of chromosomes. These changes are not inherited, but occur as random events during the formation of reproductive cells (eggs and sperm). An error in cell division called nondisjunction results in reproductive cells with an abnormal number of chromosomes. For example, a reproductive cell may accidentally gain or lose one copy of a chromosome. If one of these atypical reproductive cells contributes to the genetic makeup of a child, the child will have an extra or missing chromosome in each of the body's cells.

Changes in chromosome structure can also cause chromosomal disorders. Some changes in chromosome structure can be inherited, while others occur as random accidents during the formation of reproductive cells or in early fetal development. Because the inheritance of these changes can be complex, people concerned about this type of chromosomal abnormality may want to talk with a genetics professional.

Some cancer cells also have changes in the number or structure of their chromosomes. Because these changes occur in somatic cells (cells other than eggs and sperm), they cannot be passed from one generation to the next.

For more information about how chromosomal changes occur:

As part of its fact sheet on chromosome abnormalities, the National Human Genome Research Institute provides a discussion of how chromosome abnormalities happen. (http://www.genome.gov/11508982#6)

The Chromosome Deletion Outreach fact sheet Introduction to Chromosomes (http://chromodisorder.org/Display.aspx?ID=35) explains how structural changes occur.

The March of Dimes discusses the causes of chromosomal abnormalities in their fact sheet Chromosomal Abnormalities (http://www.marchofdimes.com/baby/chromosomal-conditions.aspx).

Additional information about how chromosomal changes happen (http://www.urmc.rochester.edu/Encyclopedia/Content.aspx?ContentTypeID=90&ContentID=P02126) is available from the University of Rochester Medical Center.

Why are some genetic conditions more common in particular ethnic groups?

Some genetic disorders are more likely to occur among people who trace their ancestry to a particular geographic area. People in an ethnic group often share certain versions of their genes, which have been passed down from common ancestors. If one of these shared genes contains a disease-causing mutation, a particular genetic disorder may be more frequently seen in the group.

Examples of genetic conditions that are more common in particular ethnic groups are sickle cell anemia, which is more common in people of African, African-American, or Mediterranean heritage; and Tay-Sachs disease, which is more likely to occur among people of Ashkenazi (eastern and central European) Jewish or French Canadian ancestry. It is important to note, however, that these disorders can occur in any ethnic group.

For more information about genetic disorders that are more common in certain groups:

The National Coalition for Health Professional Education in Genetics offers Some Frequently Asked Questions and Answers About Race, Genetics, and Healthcare (http://www.nchpeg.org/index.php?option=com_content&view=article&id= 142&Itemid=64).

Mount Sinai Hospital in Ontario, Canada provides information about conditions that are seen more commonly in people with particular ethnic backgrounds (http://www.mountsinai.on.ca/care/pdmg/genetics/ethnicity-based-conditions).

The Center for Jewish Genetics provides information on disorders that occur more frequently in people with Jewish ancestry (https://www.jewishgenetics.org/jewish-genetics) and lists genetic traits that tend to be more common in Ashkenazi Jews (https://www.jewishgenetics.org/ashkenazi-genetic-traits) and Sephardic Jews (https://www.jewishgenetics.org/sephardic-genetic-traits).

Chapter 5
Genetic Consultation

Table of Contents

What is a genetic consultation?

A genetic consultation is a health service that provides information and support to people who have, or may be at risk for, genetic disorders. During a consultation, a genetics professional meets with an individual or family to discuss genetic risks or to diagnose, confirm, or rule out a genetic condition.

Genetics professionals include medical geneticists (doctors who specialize in genetics) and genetic counselors (certified healthcare workers with experience in medical genetics and counseling). Other healthcare professionals such as nurses, psychologists, and social workers trained in genetics can also provide genetic consultations.

Consultations usually take place in a doctor's office, hospital, genetics center, or other type of medical center. These meetings are most often in-person visits with individuals or families, but they are occasionally conducted in a group or over the telephone.

For more information about genetic consultations:

MedlinePlus offers a list of links to information about genetic counseling (http://www.nlm.nih.gov/medlineplus/geneticcounseling.html).

Additional background information is provided by the National Genome Research Institute in its Frequently Asked Questions About Genetic Counseling (http://www.genome.gov/19016905).

Information about genetic counseling, including the different types of counseling, is available from the National Center for Biotechnology Information (NCBI) in the booklet Making Sense of Your Genes: A Guide to Genetic Counseling (http://www.ncbi.nlm.nih.gov/books/NBK115508/).

The Tuberous Sclerosis Alliance provides an overview of geneticists and genetic counselors (http://www.tsalliance.org/pages.aspx?content=607).

The Centre for Genetics Education also offers an introduction to genetic counseling (http://www.genetics.edu.au/Publications-and-Resources/Genetics-Fact-Sheets/FactSheet3).

Why might someone have a genetic consultation?

Individuals or families who are concerned about an inherited condition may benefit from a genetic consultation. The reasons that a person might be referred to a genetic counselor, medical geneticist, or other genetics professional include:

- A personal or family history of a genetic condition, birth defect, chromosomal disorder, or hereditary cancer.

- Two or more pregnancy losses (miscarriages), a stillbirth, or a baby who died.

- A child with a known inherited disorder, a birth defect, mental retardation, or developmental delay.

- A woman who is pregnant or plans to become pregnant at or after age 35. (Some chromosomal disorders occur more frequently in children born to older women.)

- Abnormal test results that suggest a genetic or chromosomal condition.

- An increased risk of developing or passing on a particular genetic disorder on the basis of a person's ethnic background.

- People related by blood (for example, cousins) who plan to have children together. (A child whose parents are related may be at an increased risk of inheriting certain genetic disorders.)

A genetic consultation is also an important part of the decision-making process for genetic testing. A visit with a genetics professional may be helpful even if testing is not available for a specific condition, however.

For more information about the reasons for having a genetic consultation:

An overview of indications for a genetics referral (http://www.ncbi.nlm.nih.gov/books/NBK115554/) is available from The Genetic Alliance booklet "Understanding Genetics: A Guide for Patients and Professionals."

What happens during a genetic consultation?

A genetic consultation provides information, offers support, and addresses a patient's specific questions and concerns. To help determine whether a condition has a genetic component, a genetics professional asks about a person's medical history and takes a detailed family history (a record of health information about a person's immediate and extended family). The genetics professional may also perform a physical examination and recommend appropriate tests.

If a person is diagnosed with a genetic condition, the genetics professional provides information about the diagnosis, how the condition is inherited, the chance of passing the condition to future generations, and the options for testing and treatment.

During a consultation, a genetics professional will:

- Interpret and communicate complex medical information.
- Help each person make informed, independent decisions about their health care and reproductive options.
- Respect each person's individual beliefs, traditions, and feelings.

A genetics professional will NOT:

- Tell a person which decision to make.
- Advise a couple not to have children.
- Recommend that a woman continue or end a pregnancy.
- Tell someone whether to undergo testing for a genetic disorder.

For more information about what to expect during a genetic consultation

The National Society of Genetic Counselors offers information about what to expect from a genetic counseling session as part of its FAQs About Genetic Counselors (http://nsgc.org/p/cm/ld/fid=144).

EuroGentest explains what a person can expect during a visit with a genetic specialist (http://www.eurogentest.org/index.php?id=620) and offers frequently asked questions that may be helpful during an appointment (http://www.eurogentest.org/index.php?id=615).

Information about the role of genetic counselors and the process of genetic counseling (http://www.ncbi.nlm.nih.gov/books/NBK115552/) are available from the Genetic Alliance publication "Understanding Genetics: A Guide for Patients and Professionals."

The Illinois Department of Public Health discusses genetic counseling services and provides a list of questions to ask a genetic counselor (http://www.idph.state.il.us/ HealthWellness/gencounselor.htm).

How can I find a genetics professional in my area?

To find a genetics professional in your community, you may wish to ask your doctor for a referral. If you have health insurance, you can also contact your insurance company to find a medical geneticist or genetic counselor in your area who participates in your plan.

Several resources for locating a genetics professional in your community are available online:

- The National Society of Genetic Counselors offers a searchable directory of genetic counselors in the United States and Canada (http://nsgc.org/p/cm/ld/fid=164). You can search by location, name, area of practice/specialization, and/or ZIP Code.

- The National Cancer Institute provides a Cancer Genetics Services Directory (http://www.cancer.gov/cancertopics/genetics/directory), which lists professionals who provide services related to cancer genetics. You can search by type of cancer or syndrome, location, and/or provider name.

What is the prognosis of a genetic condition?

The prognosis of a genetic condition includes its likely course, duration, and outcome. When health professionals refer to the prognosis of a disease, they may also mean the chance of recovery; however, most genetic conditions are life-long and are managed rather than cured.

Disease prognosis has multiple aspects, including:

- How long a person with the disorder is likely to live (life expectancy)
- Whether the signs and symptoms worsen (and how quickly) or are stable over time
- Quality of life, such as independence in daily activities
- Potential for complications and associated health

The prognosis of a genetic condition depends on many factors, including the specific diagnosis (http://ghr.nlm.nih.gov/handbook/consult/diagnosis) and an individual's particular signs and symptoms. Sometimes the associated genetic change, if known, can also give clues to the prognosis. Additionally, the course and outcome of a condition depends on the availability and effectiveness of treatment and management (http://ghr.nlm.nih.gov/handbook/consult/treatment) approaches. The prognosis of very rare diseases can be difficult to predict because so few affected individuals have been identified. Prognosis may also be difficult or impossible to establish if a person's diagnosis is unknown.

The prognoses of genetic disorders vary widely, often even among people with the same condition. Some genetic disorders cause physical and developmental problems that are so severe they are incompatible with life. These conditions may cause a miscarriage of an affected embryo or fetus, or an affected infant may be stillborn or die shortly after birth. People with less severe genetic conditions may live into childhood or adulthood but have a shortened lifespan due to health problems related to their disorder. Genetic conditions with a milder course may be associated with a normal lifespan and few related health issues.

The prognosis of a disease is based on probability, which means that it is likely but not certain that the disorder will follow a particular course. Your healthcare provider is the best resource for information about the prognosis of your specific genetic condition. He or she can assess your medical history and signs and symptoms to give you the most accurate estimate of your prognosis.

Learn more about the prognosis of genetic conditions:

This list of resources (http://ghr.nlm.nih.gov/handbook/consult/findingprofessional) can help you locate a genetics professional in your area.

The A.D.A.M. Medical Encyclopedia (http://www.nlm.nih.gov/medlineplus/encyclopedia.html) on MedlinePlus offers brief descriptions about many health problems, including some genetic conditions. Each page includes a section on Outlook (prognosis).

A discussion of the prognosis of disorders with a neurological basis (http://www.ninds.nih.gov/disorders/disorder_index.htm) is available from the National Institute of Neurological Disorders and Stroke (NINDS).

The National Cancer Institute (NCI) provides an overview of cancer prognosis (http://www.cancer.gov/cancertopics/factsheet/Support/prognosis-stats).

Nemours' KidsHealth has a fact sheet, When Your Baby is Born With a Health Problem (http://www.nemours.org/content/nemours/wwwv2/service/medical/geneticdisorders.html?tab=about&kidshealth=22895), that outlines what parents can expect when their infant has a genetic condition.

Local and national support and advocacy groups are also excellent resources for information about specific genetic conditions, including disease prognosis. Each condition summary (http://ghr.nlm.nih.gov/BrowseConditions) on Genetics Home Reference provides links to support and advocacy resources under the heading "Patient Support." Additionally, patient support resources related to specific genetic conditions can be identified through the Genetic Alliance's Disease InfoSearch (http://www.diseaseinfosearch.org/).

How are genetic conditions diagnosed?

A doctor may suspect a diagnosis of a genetic condition on the basis of a person's physical characteristics and family history, or on the results of a screening test.

Genetic testing is one of several tools that doctors use to diagnose genetic conditions. The approaches to making a genetic diagnosis include:

- A physical examination: Certain physical characteristics, such as distinctive facial features, can suggest the diagnosis of a genetic disorder. A geneticist will do a thorough physical examination that may include measurements such as the distance around the head (head circumference), the distance between the eyes, and the length of the arms and legs. Depending on the situation, specialized examinations such as nervous system (neurological) or eye (ophthalmologic) exams may be performed. The doctor may also use imaging studies including x-rays, computerized tomography (CT) scans, or magnetic resonance imaging (MRI) to see structures inside the body.

- Personal medical history: Information about an individual's health, often going back to birth, can provide clues to a genetic diagnosis. A personal medical history includes past health issues, hospitalizations and surgeries, allergies, medications, and the results of any medical or genetic testing that has already been done.

- Family medical history: Because genetic conditions often run in families, information about the health of family members can be a critical tool for diagnosing these disorders. A doctor or genetic counselor will ask about health conditions in an individual's parents, siblings, children, and possibly more distant relatives. This information can give clues about the diagnosis and inheritance pattern of a genetic condition in a family.

- Laboratory tests, including genetic testing: Molecular, chromosomal, and biochemical genetic testing are used to diagnose genetic disorders. Other laboratory tests that measure the levels of certain substances in blood and urine can also help suggest a diagnosis.

Genetic testing is currently available for many genetic conditions. However, some conditions do not have a genetic test; either the genetic cause of the condition is unknown or a test has not yet been developed. In these cases, a combination of the approaches listed above may be used to make a diagnosis. Even when genetic testing is available, the tools listed above are used to narrow down the possibilities (known as a differential diagnosis) and choose the most appropriate genetic tests to pursue.

A diagnosis of a genetic disorder can be made anytime during life, from before birth to old age, depending on when the features of the condition appear and the availability of testing. Sometimes, having a diagnosis can guide treatment and management decisions. A genetic diagnosis can also suggest whether other family members may be affected by or at risk of a specific disorder. Even when no treatment is available for a particular condition, having a diagnosis can help people know what to expect and may help them identify useful support and advocacy resources.

For more information about diagnosing genetic conditions:

Genetics Home Reference provides information about genetic testing (http://ghr.nlm.nih.gov/handbook/testing) and the importance of family medical history (http://ghr.nlm.nih.gov/handbook/inheritance/familyhistory). Additionally, links to information about the diagnosis of specific genetic disorders are available in each condition summary (http://ghr.nlm.nih.gov/BrowseConditions) under the heading "Where can I find information about diagnosis or management of...?"

The National Center for Biotechnology Information (NCBI) provides an in-depth guide called Understanding Genetics (http://www.ncbi.nlm.nih.gov/books/NBK132142/), which includes a chapter about how genetics professionals diagnose many types of genetic disorders.

The Centers for Disease Control and Prevention (CDC) offers a fact sheet about the diagnosis of birth defects (http://www.cdc.gov/ncbddd/birthdefects/diagnosis.html), including information about screening and diagnostic tests.

Boston Children's Hospital provides this brief overview of testing for genetic disorders (http://www.childrenshospital.org/az/Site2884/mainpageS2884P3.html).

The American College of Medical Genetics offers practice guidelines (https://www.acmg.net/ACMG/Publications/Practice_Guidelines/ACMG/Publications/Practice_Guidelines.aspx), including diagnostic criteria, for several genetic disorders. These guidelines are designed for geneticists and other healthcare providers.

The Agency for Healthcare Research and Quality's (AHRQ) National Guideline Clearinghouse has compiled screening, diagnosis, treatment, and management guidelines (http://www.guideline.gov/browse/by-topic-detail.aspx?id=19273&ct=1) for many genetic disorders.

GeneReviews (http://www.ncbi.nlm.nih.gov/books/NBK1116/), a resource from the University of Washington and the NCBI, provides detailed information about the diagnosis of specific genetic disorders as part of each peer-reviewed disease description.

How are genetic conditions treated or managed?

Many genetic disorders result from gene changes that are present in essentially every cell in the body. As a result, these disorders often affect many body systems, and most cannot be cured. However, approaches may be available to treat or manage some of the associated signs and symptoms.

For a group of genetic conditions called inborn errors of metabolism, which result from genetic changes that disrupt the production of specific enzymes, treatments sometimes include dietary changes or replacement of the particular enzyme that is missing. Limiting certain substances in the diet can help prevent the buildup of potentially toxic substances that are normally broken down by the enzyme. In some cases, enzyme replacement therapy can help compensate for the enzyme shortage. These treatments are used to manage existing signs and symptoms and may help prevent future complications.

For other genetic conditions, treatment and management strategies are designed to improve particular signs and symptoms associated with the disorder. These approaches vary by disorder and are specific to an individual's health needs. For example, a genetic disorder associated with a heart defect might be treated with surgery to repair the defect or with a heart transplant. Conditions that are characterized by defective blood cell formation, such as sickle cell disease, can sometimes be treated with a bone marrow transplant. Bone marrow transplantation can allow the formation of normal blood cells and, if done early in life, may help prevent episodes of pain and other future complications.

Some genetic changes are associated with an increased risk of future health problems, such as certain forms of cancer. One well-known example is familial breast cancer related to mutations in the BRCA1 and BRCA2 genes. Management may include more frequent cancer screening or preventive (prophylactic) surgery to remove the tissues at highest risk of becoming cancerous.

Genetic disorders may cause such severe health problems that they are incompatible with life. In the most severe cases, these conditions may cause a miscarriage of an affected embryo or fetus. In other cases, affected infants may be stillborn or die shortly after birth. Although few treatments are available for these severe genetic conditions, health professionals can often provide supportive care, such as pain relief or mechanical breathing assistance, to the affected individual.

Most treatment strategies for genetic disorders do not alter the underlying genetic mutation; however, a few disorders have been treated with gene therapy. This experimental technique involves changing a person's genes to prevent or treat a disease. Gene therapy, along with many other treatment and management approaches for genetic conditions, are under study in clinical trials.

Find out more about the treatment and management of genetic conditions:

Links to information about the treatment of specific genetic disorders are available in each Genetics Home Reference condition summary (http://ghr.nlm.nih.gov/ BrowseConditions) under the heading "Where can I find information about diagnosis or management of...?"

GeneReviews (http://www.ncbi.nlm.nih.gov/books/NBK1116/), a resource from the University of Washington and the National Center for Biotechnology Information (NCBI), provides detailed information about the management of specific genetic disorders as part of each peer-reviewed disease description.

The Agency for Healthcare Research and Quality's (AHRQ) National Guideline Clearinghouse has compiled screening, diagnosis, treatment, and management guidelines (http://www.guideline.gov/browse/by-topic.aspx) for many diseases, including genetic disorders.

Information related to the approaches discussed above is available from MedlinePlus:

- Inborn Errors of Metabolism (http://www.nlm.nih.gov/medlineplus/ency/ article/002438.htm)

- Bone Marrow Transplantation (http://www.nlm.nih.gov/medlineplus/ bonemarrowtransplantation.html)

- Palliative care (http://www.nlm.nih.gov/medlineplus/palliativecare.html) (also known as supportive care)

Genetics Home Reference offers consumer-friendly information about gene therapy (http://ghr.nlm.nih.gov/handbook/therapy), including safety, ethical issues, and availability.

ClinicalTrials.gov (http://clinicaltrials.gov/), a service of the National Institutes of Health, provides easy access to information on clinical trials. You can search for specific trials or browse by condition (http://www.clinicaltrials.gov/ct2/search/browse? brwse=cond_alpha_all),trial sponsor (http://www.clinicaltrials.gov/ct2/search/browse? brwse=spns_cat), location (http://www.clinicaltrials.gov/ct2/search/browse?brwse= locn_cat), or treatment approach (for example, drug interventions (http://www.clinicaltrials.gov/ct2/search/browse?brwse=intr_cat)).

Chapter 6

Genetic Testing

Table of Contents

What is genetic testing?

Genetic testing is a type of medical test that identifies changes in chromosomes, genes, or proteins. The results of a genetic test can confirm or rule out a suspected genetic condition or help determine a person's chance of developing or passing on a genetic disorder. More than 1,000 genetic tests are currently in use, and more are being developed.

Several methods can be used for genetic testing:

- Molecular genetic tests (or gene tests) study single genes or short lengths of DNA to identify variations or mutations that lead to a genetic disorder.

- Chromosomal genetic tests analyze whole chromosomes or long lengths of DNA to see if there are large genetic changes, such as an extra copy of a chromosome, that cause a genetic condition.

- Biochemical genetic tests study the amount or activity level of proteins; abnormalities in either can indicate changes to the DNA that result in a genetic disorder.

Genetic testing is voluntary. Because testing has benefits as well as limitations and risks, the decision about whether to be tested is a personal and complex one. A geneticist or genetic counselor can help by providing information about the pros and cons of the test and discussing the social and emotional aspects of testing.

For general information about genetic testing:

MedlinePlus offers a list of links to information about genetic testing (http://www.nlm.nih.gov/medlineplus/genetictesting.html).

The National Human Genome Research Institute provides an overview of this topic in its Frequently Asked Questions About Genetic Testing (http://www.genome.gov/19516567). Additional information about genetic testing legislation, policy, and oversight (http://www.genome.gov/10002335) is available from the Institute.

The National Institutes of Health fact sheet Genetic Testing: What It Means for Your Health and for Your Family's Health (http://www.genome.gov/Pages/Health/PatientsPublicInfo/GeneticTestingWhatItMeansForYourHealth.pdf) provides a brief overview for people considering genetic testing.

Educational resources related to genetic testing (http://geneed.nlm.nih.gov/topic_subtopic.php?tid=41&sid=42) are available from GeneEd.

The Genetics and Public Policy Center also offers information about genetic testing (http://www.dnapolicy.org/science.gt.php).

You can also search for clinical trials involving genetic testing. ClinicalTrials.gov (http://clinicaltrials.gov/), a service of the National Institutes of Health, provides easy access to information on clinical trials. You can search for specific trials or browse by condition or trial sponsor. You may wish to refer to a list of studies related to genetic testing (http://clinicaltrials.gov/search?term=%22genetic+testing%22) that are accepting (or will accept) participants.

What are the types of genetic tests?

Genetic testing can provide information about a person's genes and chromosomes. Available types of testing include:

Newborn screening

Newborn screening is used just after birth to identify genetic disorders that can be treated early in life. Millions of babies are tested each year in the United States. All states currently test infants for phenylketonuria (a genetic disorder that causes mental retardation if left untreated) and congenital hypothyroidism (a disorder of the thyroid gland). Most states also test for other genetic disorders.

Diagnostic testing

Diagnostic testing is used to identify or rule out a specific genetic or chromosomal condition. In many cases, genetic testing is used to confirm a diagnosis when a particular condition is suspected based on physical signs and symptoms. Diagnostic testing can be performed before birth or at any time during a person's life, but is not available for all genes or all genetic conditions. The results of a diagnostic test can influence a person's choices about health care and the management of the disorder.

Carrier testing

Carrier testing is used to identify people who carry one copy of a gene mutation that, when present in two copies, causes a genetic disorder. This type of testing is offered to individuals who have a family history of a genetic disorder and to people in certain ethnic groups with an increased risk of specific genetic conditions. If both parents are tested, the test can provide information about a couple's risk of having a child with a genetic condition.

Prenatal testing

Prenatal testing is used to detect changes in a fetus's genes or chromosomes before birth. This type of testing is offered during pregnancy if there is an increased risk that the baby will have a genetic or chromosomal disorder. In some cases, prenatal testing can lessen a couple's uncertainty or help them make decisions about a pregnancy. It cannot identify all possible inherited disorders and birth defects, however.

Preimplantation testing

Preimplantation testing, also called preimplantation genetic diagnosis (PGD), is a specialized technique that can reduce the risk of having a child with a particular genetic or chromosomal disorder. It is used to detect genetic changes in embryos that were created using assisted reproductive techniques such as in-vitro fertilization. In-vitro fertilization involves removing egg cells from a woman's ovaries and fertilizing them with sperm cells outside the body. To perform preimplantation testing, a small number of cells are taken from these embryos and tested for certain genetic changes. Only embryos without these changes are implanted in the uterus to initiate a pregnancy.

Predictive and presymptomatic testing

Predictive and presymptomatic types of testing are used to detect gene mutations associated with disorders that appear after birth, often later in life. These tests can be helpful to people who have a family member with a genetic disorder, but who have no features of the disorder themselves at the time of testing. Predictive testing can identify mutations that increase a person's risk of developing disorders with a genetic basis, such as certain types of cancer. Presymptomatic testing can determine whether a person will develop a genetic disorder, such as hemochromatosis (an iron overload disorder), before any signs or symptoms appear. The results of predictive and presymptomatic testing can provide information about a person's risk of developing a specific disorder and help with making decisions about medical care.

Forensic testing

Forensic testing uses DNA sequences to identify an individual for legal purposes. Unlike the tests described above, forensic testing is not used to detect gene mutations associated with disease. This type of testing can identify crime or catastrophe victims, rule out or implicate a crime suspect, or establish biological relationships between people (for example, paternity).

For more information about the uses of genetic testing:

A Brief Primer on Genetic Testing (http://www.genome.gov/10506784), which outlines the different kinds of genetic tests, is available from the National Human Genome Research Institute.

Educational resources related to patient genetic testing/carrier screening (http://geneed.nlm.nih.gov/topic_subtopic.php?tid=52&sid=55) are available from GeneEd. Johns Hopkins Medicine provides additional information about genetic carrier screening (http://www.hopkinsmedicine.org/fertility/resources/genetic_screening.html).

The Centre for Genetics Education offers an overview of prenatal testing (http://www.genetics.edu.au/Publications-and-Resources/Genetics-Fact-Sheets/FactSheet17) and fact sheets about preimplantation genetic diagnosis (http://www.genetics.edu.au/Publications-and-Resources/Genetics-Fact-Sheets/FactSheet18) and the medical applications of genetic testing and screening (http://www.genetics.edu.au/Publications-and-Resources/Genetics-Fact-Sheets/FactSheet21).

EuroGentest provides fact sheets about predictive testing (http://www.eurogentest.org/index.php?id=617) and carrier testing (http://www.eurogentest.org/index.php?id=610).

The University of Pennsylvania offers an explanation of preimplantation genetic diagnosis (http://www.pennmedicine.org/fertility/patient/clinical-services/pgd-preimplantation-genetic-diagnosis/).

Genetics Home Reference provides information and links related to newborn screening (http://ghr.nlm.nih.gov/healthTopic/newborn-screening).

The National Newborn Screening and Genetics Resource Center (http://genes-r-us.uthscsa.edu/) offers detailed information about newborn screening. Additional information about newborn screening (http://www.genetics.edu.au/Publications-and-Resources/Genetics-Fact-Sheets/FactSheet20), particularly in Australia, is available from the Centre for Genetics Education.

For information about forensic DNA testing, refer to the fact sheet about forensic genetic testing (http://www.genetics.edu.au/Publications-and-Resources/Genetics-Fact-Sheets/FactSheet22) from the Centre for Genetics Education and a page about forensic DNA analysis (http://learn.genetics.utah.edu/content/science/forensics/) from the Genetic Science Learning Center at the Univeristy of Utah.

How is genetic testing done?

Once a person decides to proceed with genetic testing, a medical geneticist, primary care doctor, specialist, or nurse practitioner can order the test. Genetic testing is often done as part of a genetic consultation.

Genetic tests are performed on a sample of blood, hair, skin, amniotic fluid (the fluid that surrounds a fetus during pregnancy), or other tissue. For example, a procedure called a buccal smear uses a small brush or cotton swab to collect a sample of cells from the inside surface of the cheek. The sample is sent to a laboratory where technicians look for specific changes in chromosomes, DNA, or proteins, depending on the suspected disorder. The laboratory reports the test results in writing to a person's doctor or genetic counselor, or directly to the patient if requested.

Newborn screening tests are done on a small blood sample, which is taken by pricking the baby's heel. Unlike other types of genetic testing, a parent will usually only receive the result if it is positive. If the test result is positive, additional testing is needed to determine whether the baby has a genetic disorder.

Before a person has a genetic test, it is important that he or she understands the testing procedure, the benefits and limitations of the test, and the possible consequences of the test results. The process of educating a person about the test and obtaining permission is called informed consent (http://ghr.nlm.nih.gov/handbook/testing/informedconsent).

For more information about genetic testing procedures:

Scientific Testimony, an online journal, provides an introduction to DNA testing techniques (http://www.scientific.org/tutorials/articles/riley/riley.html) written for the general public.

The National Genetics and Genomics Education Centre of the National Health Service (UK) offers an overview of what happens in the laboratory (http://www.geneticseducation.nhs.uk/laboratory-process-and-testing-techniques) during genetic testing.

The Genetic Science Learning Center at the University of Utah provides an interactive animation of DNA extraction techniques (http://learn.genetics.utah.edu/content/labs/extraction/).

What is informed consent?

Before a person has a genetic test, it is important that he or she fully understands the testing procedure, the benefits and limitations of the test, and the possible consequences of the test results. The process of educating a person about the test and obtaining permission to carry out testing is called informed consent. "Informed" means that the person has enough information to make an educated decision about testing; "consent" refers to a person's voluntary agreement to have the test done.

In general, informed consent can only be given by adults who are competent to make medical decisions for themselves. For children and others who are unable to make their own medical decisions (such as people with impaired mental status), informed consent can be given by a parent, guardian, or other person legally responsible for making decisions on that person's behalf.

Informed consent for genetic testing is generally obtained by a doctor or genetic counselor during an office visit. The healthcare provider will discuss the test and answer any questions. If the person wishes to have the test, he or she will then usually read and sign a consent form.

Several factors are commonly included on an informed consent form:

- A general description of the test, including the purpose of the test and the condition for which the testing is being performed.

- How the test will be carried out (http://ghr.nlm.nih.gov/handbook/testing/procedure) (for example, a blood sample).

- What the test results mean (http://ghr.nlm.nih.gov/handbook/testing/interpretingresults), including positive and negative results, and the potential for uninformative results or incorrect results such as false positives or false negatives.

- Any physical or emotional risks associated with the test (http://ghr.nlm.nih.gov/handbook/testing/riskslimitations).

- Whether the results can be used for research purposes (http://ghr.nlm.nih.gov/handbook/testing/researchtesting).

- Whether the results might provide information about other family members' health, including the risk of developing a particular condition or the possibility of having affected children.

- How and to whom test results will be reported and under what circumstances results can be disclosed (for example, to health insurance providers).

- What will happen to the test specimen after the test is complete.

- Acknowledgement that the person requesting testing has had the opportunity to discuss the test with a healthcare professional.

- The individual's signature, and possibly that of a witness.

The elements of informed consent may vary, because some states have laws that specify factors that must be included. (For example, some states require disclosure that the test specimen will be destroyed within a certain period of time after the test is complete.)

Informed consent is not a contract, so a person can change his or her mind at any time after giving initial consent. A person may choose not to go through with genetic testing even after the test sample has been collected. A person simply needs to notify the healthcare provider if he or she decides not to continue with the testing process.

For more information about informed consent:

MedlinePlus offers general information about informed consent by adults (http://www.nlm.nih.gov/medlineplus/ency/patientinstructions/000445.htm).

The National Cancer Institute discusses informed consent for genetic testing in the context of inherited cancer syndromes (http://www.cancer.gov/cancertopics/pdq/genetics/risk-assessment-and-counseling/HealthProfessional/page5#Section_226).

The National Human Genome Research Institute provides information about policies and legislation related to informed consent for genetic research studies and testing (http://www.genome.gov/10002332).

The Centers for Disease Control and Prevention offers several examples of state-required components of informed consent for genetic testing (http://www.cdc.gov/mmwr/preview/mmwrhtml/rr5806a3.htm).

Additional information about informed consent (http://www.phgfoundation.org/tutorials/informedConsent/) is available in a tutorial from the PHG Foundation (UK).

What is direct-to-consumer genetic testing?

Traditionally, genetic tests have been available only through healthcare providers such as physicians, nurse practitioners, and genetic counselors. Healthcare providers order the appropriate test from a laboratory, collect and send the samples, and interpret the test results. Direct-to-consumer genetic testing refers to genetic tests that are marketed directly to consumers via television, print advertisements, or the Internet. This form of testing, which is also known as at-home genetic testing, provides access to a person's genetic information without necessarily involving a doctor or insurance company in the process.

If a consumer chooses to purchase a genetic test directly, the test kit is mailed to the consumer instead of being ordered through a doctor's office. The test typically involves collecting a DNA sample at home, often by swabbing the inside of the cheek, and mailing the sample back to the laboratory. In some cases, the person must visit a health clinic to have blood drawn. Consumers are notified of their results by mail or over the telephone, or the results are posted online. In some cases, a genetic counselor or other healthcare provider is available to explain the results and answer questions. The price for this type of at-home genetic testing ranges from several hundred dollars to more than a thousand dollars.

The growing market for direct-to-consumer genetic testing may promote awareness of genetic diseases, allow consumers to take a more proactive role in their health care, and offer a means for people to learn about their ancestral origins. At-home genetic tests, however, have significant risks and limitations. Consumers are vulnerable to being misled by the results of unproven or invalid tests. Without guidance from a healthcare provider, they may make important decisions about treatment or prevention based on inaccurate, incomplete, or misunderstood information about their health. Consumers may also experience an invasion of genetic privacy if testing companies use their genetic information in an unauthorized way.

Genetic testing provides only one piece of information about a person's health—other genetic and environmental factors, lifestyle choices, and family medical history also affect a person's risk of developing many disorders. These factors are discussed during a consultation with a doctor or genetic counselor, but in many cases are not addressed by at-home genetic tests. More research is needed to fully understand the benefits and limitations of direct-to-consumer genetic testing.

For more information about direct-to-consumer genetic testing:

The American College of Medical Genetics, which is a national association of doctors specializing in genetics, has issued a statement on direct-to-consumer genetic testing (https://www.acmg.net/StaticContent/StaticPages/Direct_Consumer.pdf).

The American Society of Human Genetics, a professional membership organization for specialists in genetics, has also issued a statement on direct-to-consumer genetic testing in the United States (http://ashg.org/pdf/dtc_statement.pdf).

The Federal Trade Commission (FTC) works to protect consumers and promote truth in advertising. The FTC offers an overview of direct-to-consumer genetic testing (http://www.consumer.ftc.gov/articles/0166-direct-consumer-genetic-tests) and a fact sheet for older people (http://www.consumer.ftc.gov/articles/0325-home-genetic-tests-health-information-older-people) about the benefits and risks of at-home genetic tests.

An issue brief on direct-to-consumer genetic testing (http://www.dnapolicy.org/policy.issue.php?action=detail&issuebrief_id=32) is available from the Genetics & Public Policy Center.

The Genetic Alliance also provides information about genetic testing (http://www.geneticalliance.org/advocacy/policyissues/genetictesting), including issues surrounding direct-to-consumer genetic testing.

Additional information about direct-to-consumer marketing of genetic tests (http://www.genome.gov/12010659) is available from the National Human Genome Research Institute.

EuroGentest offers a list of publications related to direct-to-consumer genetic testing in Europe (http://www.eurogentest.org/index.php?id=678).

How can consumers be sure a genetic test is valid and useful?

Before undergoing genetic testing, it is important to be sure that the test is valid and useful. A genetic test is valid if it provides an accurate result. Two main measures of accuracy apply to genetic tests: analytical validity and clinical validity. Another measure of the quality of a genetic test is its usefulness, or clinical utility.

- Analytical validity refers to how well the test predicts the presence or absence of a particular gene or genetic change. In other words, can the test accurately detect whether a specific genetic variant is present or absent?

- Clinical validity refers to how well the genetic variant being analyzed is related to the presence, absence, or risk of a specific disease.

- Clinical utility refers to whether the test can provide information about diagnosis, treatment, management, or prevention of a disease that will be helpful to a consumer.

All laboratories that perform health-related testing, including genetic testing, are subject to federal regulatory standards called the Clinical Laboratory Improvement Amendments (CLIA) or even stricter state requirements. CLIA standards cover how tests are performed, the qualifications of laboratory personnel, and quality control and testing procedures for each laboratory. By controlling the quality of laboratory practices, CLIA standards are designed to ensure the analytical validity of genetic tests.

CLIA standards do not address the clinical validity or clinical utility of genetic tests. The Food and Drug Administration (FDA) requires information about clinical validity for some genetic tests. Additionally, the state of New York requires information on clinical validity for all laboratory tests performed for people living in that state. Consumers, health providers, and health insurance companies are often the ones who determine the clinical utility of a genetic test.

It can be difficult to determine the quality of a genetic test sold directly to the public. Some providers of direct-to-consumer genetic tests are not CLIA-certified, so it can be difficult to tell whether their tests are valid. If providers of direct-to-consumer genetic tests offer easy-to-understand information about the scientific basis of their tests, it can help consumers make more informed decisions. It may also be helpful to discuss any concerns with a health professional before ordering a direct-to-consumer genetic test.

For more information about determining the quality of genetic tests:

The Centers for Disease Control and Prevention (CDC) provides an explanation of the factors used to evaluate genetic tests (http://www.cdc.gov/genomics/gtesting/ACCE/index.htm), including analytical validity, clinical validity, and clinical utility, as part of their ACCE project. Additional information about the ACCE framework (http://www.phgfoundation.org/tutorials/acce/) is available in an interactive tutorial from the PHG Foundation.

A brief overview of the regulation of genetic testing (http://www.dnapolicy.org/policy.issue.php?action=detail&issuebrief_id=10) is available from the Genetics & Public Policy Center.

The Genetic Alliance offers information about the quality of genetic tests and current public policy issues (http://www.geneticalliance.org/advocacy/policyissues/genetictesting) surrounding their regulation.

Interactive tutorials about analytical validity (http://www.phgfoundation.org/tutorials/acce/2.html), clinical validity (http://www.phgfoundation.org/tutorials/acce/3.html), and clinical utility (http://www.phgfoundation.org/tutorials/acce/5.html) are available from the PHG Foundation.

The World Health Organization discusses quality and safety in genetic testing (http://www.who.int/genomics/policy/quality_safety/en/index1.html).

The U.S. Centers for Medicare and Medicaid Services (CMS) provide an overview of the Clinical Laboratory Improvement Amendments (CLIA) (http://www.cms.gov/Regulations-and-Guidance/Legislation/CLIA/).

What do the results of genetic tests mean?

The results of genetic tests are not always straightforward, which often makes them challenging to interpret and explain. Therefore, it is important for patients and their families to ask questions about the potential meaning of genetic test results both before and after the test is performed. When interpreting test results, healthcare professionals consider a person's medical history, family history, and the type of genetic test that was done.

A positive test result means that the laboratory found a change in a particular gene, chromosome, or protein of interest. Depending on the purpose of the test, this result may confirm a diagnosis, indicate that a person is a carrier of a particular genetic mutation, identify an increased risk of developing a disease (such as cancer) in the future, or suggest a need for further testing. Because family members have some genetic material in common, a positive test result may also have implications for certain blood relatives of the person undergoing testing. It is important to note that a positive result of a predictive or presymptomatic genetic test usually cannot establish the exact risk of developing a disorder. Also, health professionals typically cannot use a positive test result to predict the course or severity of a condition.

A negative test result means that the laboratory did not find a change in the gene, chromosome, or protein under consideration. This result can indicate that a person is not affected by a particular disorder, is not a carrier of a specific genetic mutation, or does not have an increased risk of developing a certain disease. It is possible, however, that the test missed a disease-causing genetic alteration because many tests cannot detect all genetic changes that can cause a particular disorder. Further testing may be required to confirm a negative result.

In some cases, a negative result might not give any useful information. This type of result is called uninformative, indeterminate, inconclusive, or ambiguous. Uninformative test results sometimes occur because everyone has common, natural variations in their DNA, called polymorphisms, that do not affect health. If a genetic test finds a change in DNA that has not been associated with a disorder in other people, it can be difficult to tell whether it is a natural polymorphism or a disease-causing mutation. An uninformative result cannot confirm or rule out a specific diagnosis, and it cannot indicate whether a person has an increased risk of developing a disorder. In some cases, testing other affected and unaffected family members can help clarify this type of result.

For more information about interpreting genetic test results:

The National Cancer Institute fact sheet Genetic Testing for Hereditary Cancer Syndromes (http://www.cancer.gov/cancertopics/factsheet/Risk/genetic-testing)

provides an explanation of positive and negative genetic test results. (Scroll down to question 6, "What do the results of genetic testing mean?")

The National Coalition for Health Professional Education in Genetics (NCHPEG) offers more information in their fact sheet Interpreting the Results of a Genetic or Genomic Test (http://www.nchpeg.org/index.php?option=com_content&view=article&id=172&Itemid=64).

The National Women's Health Resource Center offers a list of questions about genetic testing (http://www.healthywomen.org/condition/genetic-testing#hc-tab-1), including the meaning of test results, that patients and families can ask their healthcare professional.

Additional information about understanding genetic test results (http://www.geneticseducation.nhs.uk/understanding-genetic-test-reports) is available from the National Genetics and Genomics Education Centre of the National Health Service (UK).

What is the cost of genetic testing, and how long does it take to get the results?

The cost of genetic testing can range from under $100 to more than $2,000, depending on the nature and complexity of the test. The cost increases if more than one test is necessary or if multiple family members must be tested to obtain a meaningful result. For newborn screening, costs vary by state. Some states cover part of the total cost, but most charge a fee of $15 to $60 per infant.

From the date that a sample is taken, it may take a few weeks to several months to receive the test results. Results for prenatal testing are usually available more quickly because time is an important consideration in making decisions about a pregnancy. The doctor or genetic counselor who orders a particular test can provide specific information about the cost and time frame associated with that test.

For more information about the logistics of genetic testing:

EuroGentest offers a fact sheet about genetic testing laboratories (http://www.eurogentest.org/index.php?id=621), including the reasons why some genetic test results take longer than others.

Will health insurance cover the costs of genetic testing?

In many cases, health insurance plans will cover the costs of genetic testing when it is recommended by a person's doctor. Health insurance providers have different policies about which tests are covered, however. A person interested in submitting the costs of testing may wish to contact his or her insurance company beforehand to ask about coverage.

Some people may choose not to use their insurance to pay for testing because the results of a genetic test can affect a person's health insurance coverage. Instead, they may opt to pay out-of-pocket for the test. People considering genetic testing may want to find out more about their state's privacy protection laws before they ask their insurance company to cover the costs. (Refer to What is genetic discrimination? (http://ghr.nlm.nih.gov/handbook/testing/discrimination) for more information.)

For more information about insurance coverage of genetic testing:

The National Human Genome Research Institute provides information about Coverage and Reimbursement of Genetic Tests (http://www.genome.gov/19016729).

Genes In Life discusses insurance coverage (http://www.genesinlife.org/after-diagnosis/plan-future/insurance-and-financial-planning) and reimbursement (http://www.genesinlife.org/testing-services/testing-genetic-conditions/reimbursement-genetic-testing) for genetic testing.

What are the benefits of genetic testing?

Genetic testing has potential benefits whether the results are positive or negative for a gene mutation. Test results can provide a sense of relief from uncertainty and help people make informed decisions about managing their health care. For example, a negative result can eliminate the need for unnecessary checkups and screening tests in some cases. A positive result can direct a person toward available prevention, monitoring, and treatment options. Some test results can also help people make decisions about having children. Newborn screening can identify genetic disorders early in life so treatment can be started as early as possible.

For more information about the benefits of genetic testing:

EuroGentest offers a fact sheet about genetic testing (http://www.eurogentest.org/index.php?id=622), including a section on its benefits.

Additional information about the potential benefits of genetic testing (http://www.ucdenver.edu/academics/colleges/medicalschool/programs/Adult%20Medical%20Genetics/GeneticTestingInfo/Pages/GeneticTestingInfo.aspx#tab-2) is available from the University of Colorado.

What are the risks and limitations of genetic testing?

The physical risks associated with most genetic tests are very small, particularly for those tests that require only a blood sample or buccal smear (a procedure that samples cells from the inside surface of the cheek). The procedures used for prenatal testing carry a small but real risk of losing the pregnancy (miscarriage) because they require a sample of amniotic fluid or tissue from around the fetus.

Many of the risks associated with genetic testing involve the emotional, social, or financial consequences of the test results. People may feel angry, depressed, anxious, or guilty about their results. In some cases, genetic testing creates tension within a family because the results can reveal information about other family members in addition to the person who is tested. The possibility of genetic discrimination in employment or insurance is also a concern. (Refer to What is genetic discrimination? (http://ghr.nlm.nih.gov/handbook/testing/discrimination) for additional information.)

Genetic testing can provide only limited information about an inherited condition. The test often can't determine if a person will show symptoms of a disorder, how severe the symptoms will be, or whether the disorder will progress over time. Another major limitation is the lack of treatment strategies for many genetic disorders once they are diagnosed.

A genetics professional can explain in detail the benefits, risks, and limitations of a particular test. It is important that any person who is considering genetic testing understand and weigh these factors before making a decision.

For more information about the risks and limitations of genetic testing:

The American College of Medical Genetics and Genomics (ACMG) published a policy statement about the risks associated with incorrect genetic test results or interpretation (http://www.acmg.net/docs/LDT_Release.pdf).

EuroGentest offers a fact sheet about genetic testing (http://www.eurogentest.org/index.php?id=622), including a section on its possible risks and limitations.

Additional information about the risks and limitations of genetic testing (http://www.ucdenver.edu/academics/colleges/medicalschool/programs/Adult%20 Medical%20Genetics/GeneticTestingInfo/Pages/GeneticTestingInfo.aspx#tab-3) is available from the University of Colorado.

What is genetic discrimination?

Genetic discrimination occurs when people are treated differently by their employer or insurance company because they have a gene mutation that causes or increases the risk of an inherited disorder. Fear of discrimination is a common concern among people considering genetic testing.

Several laws at the federal and state levels help protect people against genetic discrimination. In particular, a federal law called the Genetic Information Nondiscrimination Act (GINA) is designed to protect people from this form of discrimination.

GINA has two parts: Title I, which prohibits genetic discrimination in health insurance, and Title II, which prohibits genetic discrimination in employment. Title I makes it illegal for health insurance providers to use or require genetic information to make decisions about a person's insurance eligibility or coverage. This part of the law went into effect on May 21, 2009. Title II makes it illegal for employers to use a person's genetic information when making decisions about hiring, promotion, and several other terms of employment. This part of the law went into effect on November 21, 2009.

GINA and other laws do not protect people from genetic discrimination in every circumstance. For example, GINA does not apply when an employer has fewer than 15 employees. It does not cover people in the U.S. military or those receiving health benefits through the Veterans Health Administration or Indian Health Service. GINA also does not protect against genetic discrimination in forms of insurance other than health insurance, such as life, disability, or long-term care insurance.

For more information about genetic discrimination and GINA:

The National Human Genome Research Institute provides a detailed discussion of genetic discrimination and current laws that address this issue:

- Genetic Discrimination (http://www.genome.gov/10002077)

- NHGRI Genome Statute and Legislation Database (http://www.genome.gov/PolicyEthics/LegDatabase/pubsearch.cfm)

- Genetic Information Nondiscrimination Act (GINA) of 2008 (http://www.genome.gov/24519851)

The Genetic Alliance offers links to resources and policy statements on genetic discrimination (http://www.geneticalliance.org/advocacy/policyissues/geneticdiscrimination).

The Smithsonian National Museum of Natural History's exhibit 'Genome: Unlocking Life's Code' discusses GINA's implementation (http://unlockinglifescode.org/explore/genomics-society/feature-story-gina-protection-misuse-genetic-information).

More detailed information about GINA is available from these resources:

- Genetics & Public Policy Center (http://www.dnapolicy.org/resources/WhatGINAdoesanddoesnotdochart.pdf)

- Coalition for Genetic Fairness (http://www.geneticfairness.org/ginaresource.html)

- GINAHelp.org (http://www.ginahelp.org/)

The U.S Department of Energy Human Genome Project highlights major pieces of genetics legislation (http://web.ornl.gov/sci/techresources/Human_Genome/elsi/legislat.shtml), including sections on federal policy history, state policy history, and how federal anti-discrimination laws before GINA apply to genetics.

Can genes be patented?

A gene patent is the exclusive rights to a specific sequence of DNA (a gene) given by a government to the individual, organization, or corporation who claims to have first identified the gene. Once granted a gene patent, the holder of the patent dictates how the gene can be used, in both commercial settings, such as clinical genetic testing, and in noncommercial settings, including research, for 20 years from the date of the patent. Gene patents have often resulted in companies having sole ownership of genetic testing for patented genes.

On June 13, 2013, in the case of the Association for Molecular Pathology v. Myriad Genetics, Inc., the Supreme Court of the United States ruled that human genes cannot be patented in the U.S. because DNA is a "product of nature." The Court decided that because nothing new is created when discovering a gene, there is no intellectual property to protect, so patents cannot be granted. Prior to this ruling, more than 4,300 human genes were patented. The Supreme Court's decision invalidated those gene patents, making the genes accessible for research and for commercial genetic testing.

The Supreme Court's ruling did allow that DNA manipulated in a lab is eligible to be patented because DNA sequences altered by humans are not found in nature. The Court specifically mentioned the ability to patent a type of DNA known as complementary DNA (cDNA). This synthetic DNA is produced from the molecule that serves as the instructions for making proteins (called messenger RNA).

For more information about gene patenting and the Supreme Court ruling:

Read the Supreme Court ruling (http://www.supremecourt.gov/opinions/12pdf/12-398_1b7d.pdf) against gene patenting.

The National Institutes of Health (http://www.nih.gov/about/director/06132013_statement_genepatent.htm), the American College of Medical Genetics and Genomics (https://www.acmg.net/docs/GenePatientsDecision.pdf), and the American Medical Association (http://www.ama-assn.org/ama/pub/news/news/2013/2013-06-13-end-to-human-gene-patents.page) voice their support for the Supreme Court's ruling on gene patents.

The Genetics & Public Policy Center at Johns Hopkins University summarizes the impact (http://www.dnapolicy.org/news.release.php?action=detail&pressrelease_id=153) of the Supreme Court's ruling on gene patenting.

The National Human Genome Research Institute discusses the relationship between Intellectual Property and Genomics (http://www.genome.gov/19016590)

How does genetic testing in a research setting differ from clinical genetic testing?

The main differences between clinical genetic testing and research testing are the purpose of the test and who receives the results. The goals of research testing include finding unknown genes, learning how genes work, developing tests for future clinical use, and advancing our understanding of genetic conditions. The results of testing done as part of a research study are usually not available to patients or their healthcare providers. Clinical testing, on the other hand, is done to find out about an inherited disorder in an individual patient or family. People receive the results of a clinical test and can use them to help them make decisions about medical care or reproductive issues.

It is important for people considering genetic testing to know whether the test is available on a clinical or research basis. Clinical and research testing both involve a process of informed consent (http://ghr.nlm.nih.gov/handbook/testing/informedconsent) in which patients learn about the testing procedure, the risks and benefits of the test, and the potential consequences of testing.

For more information about the differences between clinical and research testing:

The Ohio State University's Wexner Medical Center describes the difference between clinical and research genetic testing (http://internalmedicine.osu.edu/genetics/patient-care/genetic-testing-facts/index.cfm).

The Sudden Arrhythmia Death Syndromes (SADS) Foundation also outlines the major differences between clinical tests and research tests (http://www.sads.org/Living-with-SADS/Genetic-Testing/Genetic-Testing---Clinical-vs--Research).

The Columbia University Medical Center offers a table that summarizes the major differences between clinical genetic testing and genetic research (http://cpmcnet.columbia.edu/dept/neurology/movdis/genetic/genetic-research.html).

Additional information about clinical and research tests (http://www.ncbi.nlm.nih.gov/gtr/docs/about/#tests) is available from the Genetic Testing Registry.

What is genetic ancestry testing?

Genetic ancestry testing, or genetic genealogy, is a way for people interested in family history (genealogy) to go beyond what they can learn from relatives or from historical documentation. Examination of DNA variations can provide clues about where a person's ancestors might have come from and about relationships between families. Certain patterns of genetic variation are often shared among people of particular backgrounds. The more closely related two individuals, families, or populations are, the more patterns of variation they typically share.

Three types of genetic ancestry testing are commonly used for genealogy:

- Y chromosome testing: Variations in the Y chromosome, passed exclusively from father to son, can be used to explore ancestry in the direct male line. Y chromosome testing can only be done on males, because females do not have a Y chromosome. However, women interested in this type of genetic testing sometimes recruit a male relative to have the test done. Because the Y chromosome is passed on in the same pattern as are family names in many cultures, Y chromosome testing is often used to investigate questions such as whether two families with the same surname are related.

- Mitochondrial DNA testing: This type of testing identifies genetic variations in mitochondrial DNA. Although most DNA is packaged in chromosomes within the cell nucleus, cell structures called mitochondria also have a small amount of their own DNA (known as mitochondrial DNA). Both males and females have mitochondrial DNA, which is passed on from their mothers, so this type of testing can be used by either sex. It provides information about the direct female ancestral line. Mitochondrial DNA testing can be useful for genealogy because it preserves information about female ancestors that may be lost from the historical record because of the way surnames are often passed down.

- Single nucleotide polymorphism (http://ghr.nlm.nih.gov/handbook/genomicresearch/snp) testing: These tests evaluate large numbers of variations (single nucleotide polymorphisms or SNPs) across a person's entire genome. The results are compared with those of others who have taken the tests to provide an estimate of a person's ethnic background. For example, the pattern of SNPs might indicate that a person's ancestry is approximately 50 percent African, 25 percent European, 20 percent Asian, and 5 percent unknown. Genealogists use this type of test because Y chromosome and mitochondrial DNA test results, which represent only single ancestral lines, do not capture the overall ethnic background of an individual.

Genetic ancestry testing has a number of limitations. Test providers compare individuals' test results to different databases of previous tests, so ethnicity estimates may not be consistent from one provider to another. Also, because most human populations have migrated many times throughout their history and mixed with nearby groups, ethnicity estimates based on genetic testing may differ from an individual's expectations. In ethnic groups with a smaller range of genetic variation due to the group's size and history, most members share many SNPs, and it may be difficult to distinguish people who have a relatively recent common ancestor, such as fourth cousins, from the group as a whole.

Genetic ancestry testing is offered by several companies and organizations. Most companies provide online forums and other services to allow people who have been tested to share and discuss their results with others, which may allow them to discover previously unknown relationships. On a larger scale, combined genetic ancestry test results from many people can be used by scientists to explore the history of populations as they arose, migrated, and mixed with other groups.

For more information about genetic ancestry testing:

The University of Utah provides video tutorials (http://learn.genetics.utah.edu/content/chromosomes/molgen/) on molecular genealogy.

The International Society of Genetic Genealogy (http://www.isogg.org/) promotes the use of DNA testing in genealogy.

The American Society of Human Genetics (ASHG) developed a position paper on ancestry testing (http://www.ashg.org/pdf/ASHGAncestryTestingStatement_FINAL.pdf).

Detailed information about genetic ancestry testing (http://www.senseaboutscience.org/data/files/resources/119/Sense-About-Genetic-Ancestry-Testing.pdf) is available from Sense About Science.

The Smithsonian National Museum of Natural History's exhibit 'Genome: Unlocking Life's Code' discusses genetic ancestry testing (http://unlockinglifescode.org/explore/our-genomic-journey/our-origins-and-ancestry). The exhibit also discusses the African Diaspora (http://unlockinglifescode.org/explore/our-genomic-journey/feature-story-african-diaspora) and its influence on heredity and genealogy.

Chapter 7
Gene Therapy

Table of Contents

What is gene therapy?

Gene therapy is an experimental technique that uses genes to treat or prevent disease. In the future, this technique may allow doctors to treat a disorder by inserting a gene into a patient's cells instead of using drugs or surgery. Researchers are testing several approaches to gene therapy, including:

- Replacing a mutated gene that causes disease with a healthy copy of the gene.

- Inactivating, or "knocking out," a mutated gene that is functioning improperly.

- Introducing a new gene into the body to help fight a disease.

Although gene therapy is a promising treatment option for a number of diseases (including inherited disorders, some types of cancer, and certain viral infections), the technique remains risky and is still under study to make sure that it will be safe and effective. Gene therapy is currently only being tested for the treatment of diseases that have no other cures.

For general information about gene therapy:

MedlinePlus from the National Library of Medicine offers a list of links to information about genes and gene therapy (http://www.nlm.nih.gov/medlineplus/genesandgenetherapy.html).

Educational resources related to gene therapy (http://geneed.nlm.nih.gov/topic_subtopic.php?tid=142&sid=144) are available from GeneEd.

The Genetic Science Learning Center at the University of Utah provides an interactive introduction to gene therapy (http://learn.genetics.utah.edu/content/genetherapy/gtintro/) and a discussion of several diseases for which gene therapy has been successful (http://learn.genetics.utah.edu/content/genetherapy/gtsuccess/).

The Centre for Genetics Education provides an introduction to gene therapy (http://www.genetics.edu.au/Publications-and-Resources/Genetics-Fact-Sheets/FactSheet27), including a discussion of ethical and safety considerations.

KidsHealth from Nemours offers a fact sheet called Gene Therapy and Children (http://kidshealth.org/parent/system/medical/gene_therapy.html).

Additional information about gene therapy (http://www.geneticseducation.nhs.uk/genomics-in-health/developing-new-therapies) is available from the National Genetics and Genomics Education Centre of the National Health Service (UK)

How does gene therapy work?

Gene therapy is designed to introduce genetic material into cells to compensate for abnormal genes or to make a beneficial protein. If a mutated gene causes a necessary protein to be faulty or missing, gene therapy may be able to introduce a normal copy of the gene to restore the function of the protein.

A gene that is inserted directly into a cell usually does not function. Instead, a carrier called a vector is genetically engineered to deliver the gene. Certain viruses are often used as vectors because they can deliver the new gene by infecting the cell. The viruses are modified so they can't cause disease when used in people. Some types of virus, such as retroviruses, integrate their genetic material (including the new gene) into a chromosome in the human cell. Other viruses, such as adenoviruses, introduce their DNA into the nucleus of the cell, but the DNA is not integrated into a chromosome.

The vector can be injected or given intravenously (by IV) directly into a specific tissue in the body, where it is taken up by individual cells. Alternately, a sample of the patient's cells can be removed and exposed to the vector in a laboratory setting. The cells containing the vector are then returned to the patient. If the treatment is successful, the new gene delivered by the vector will make a functioning protein.

Researchers must overcome many technical challenges before gene therapy will be a practical approach to treating disease. For example, scientists must find better ways to deliver genes and target them to particular cells. They must also ensure that new genes are precisely controlled by the body.

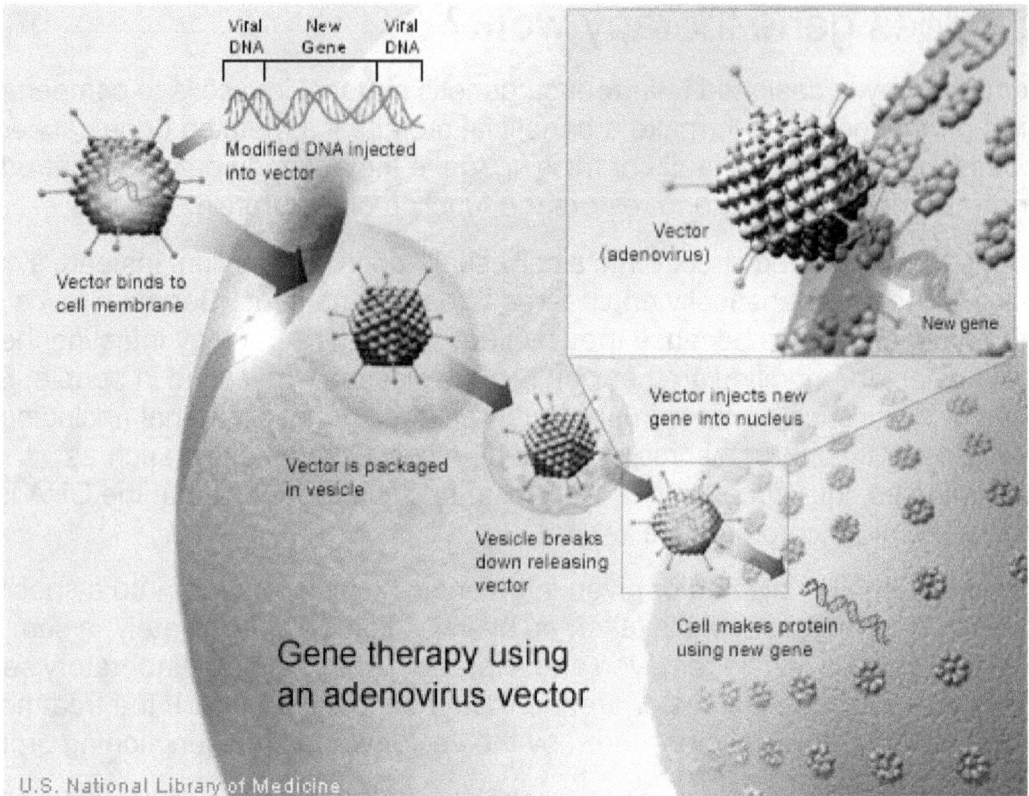

Vital DNA | New Gene | Vital DNA

Modified DNA injected into vector

Vector binds to cell membrane

Vector (adenovirus)

New gene

Vector is packaged in vesicle

Vector injects new gene into nucleus

Vesicle breaks down releasing vector

Cell makes protein using new gene

Gene therapy using an adenovirus vector

U.S. National Library of Medicine

A new gene is injected into an adenovirus vector, which is used to introduce the modified DNA into a human cell. If the treatment is successful, the new gene will make a functional protein.

For more information about how gene therapy works:

The Genetic Science Learning Center at the University of Utah provides information about various technical aspects of gene therapy in Gene Delivery: Tools of the Trade (http://learn.genetics.utah.edu/content/genetherapy/gttools/). They also discuss other approaches to gene therapy (http://learn.genetics.utah.edu/content/genetherapy/gtapproaches/) and offer a related learning activity called Space Doctor (http://learn.genetics.utah.edu/content/genetherapy/gtdoctor/).

The Better Health Channel from the State Government of Victoria (Australia) provides a brief introduction to gene therapy (http://www.betterhealth.vic.gov.au/bhcv2/bhcarticles.nsf/pages/Gene_therapy), including the gene therapy process and delivery techniques.

Penn Medicine's Oncolink describes how gene therapy works and how it is administered to patients (http://www.oncolink.org/treatment/article.cfm?aid=585&id=323&c=15).

Is gene therapy safe?

Gene therapy is under study to determine whether it could be used to treat disease. Current research is evaluating the safety of gene therapy; future studies will test whether it is an effective treatment option. Several studies have already shown that this approach can have very serious health risks, such as toxicity, inflammation, and cancer. Because the techniques are relatively new, some of the risks may be unpredictable; however, medical researchers, institutions, and regulatory agencies are working to ensure that gene therapy research is as safe as possible.

Comprehensive federal laws, regulations, and guidelines help protect people who participate in research studies (called clinical trials). The U.S. Food and Drug Administration (FDA) regulates all gene therapy products in the United States and oversees research in this area. Researchers who wish to test an approach in a clinical trial must first obtain permission from the FDA. The FDA has the authority to reject or suspend clinical trials that are suspected of being unsafe for participants.

The National Institutes of Health (NIH) also plays an important role in ensuring the safety of gene therapy research. NIH provides guidelines for investigators and institutions (such as universities and hospitals) to follow when conducting clinical trials with gene therapy. These guidelines state that clinical trials at institutions receiving NIH funding for this type of research must be registered with the NIH Office of Biotechnology Activities. The protocol, or plan, for each clinical trial is then reviewed by the NIH Recombinant DNA Advisory Committee (RAC) to determine whether it raises medical, ethical, or safety issues that warrant further discussion at one of the RAC's public meetings.

An Institutional Review Board (IRB) and an Institutional Biosafety Committee (IBC) must approve each gene therapy clinical trial before it can be carried out. An IRB is a committee of scientific and medical advisors and consumers that reviews all research within an institution. An IBC is a group that reviews and approves an institution's potentially hazardous research studies. Multiple levels of evaluation and oversight ensure that safety concerns are a top priority in the planning and carrying out of gene therapy research.

For more information about the safety and oversight of gene therapy:

Information about the development of new gene therapies and the FDA's role in overseeing the safety of gene therapy research can be found in the fact sheet Human Gene Therapies: Novel Product Development Q&A (http://www.fda.gov/ForConsumers/ConsumerUpdates/ucm103331.htm).

The Genetic Science Learning Center at the University of Utah explains challenges related to gene therapy (http://learn.genetics.utah.edu/content/genetherapy/gtchallenges/).

The NIH's Office of Biotechnology Activities provides NIH guidelines for biosafety (http://osp.od.nih.gov/office-biotechnology-activities/biosafety/nih-guidelines).

What are the ethical issues surrounding gene therapy?

Because gene therapy involves making changes to the body's set of basic instructions, it raises many unique ethical concerns. The ethical questions surrounding gene therapy include:

- How can "good" and "bad" uses of gene therapy be distinguished?

- Who decides which traits are normal and which constitute a disability or disorder?

- Will the high costs of gene therapy make it available only to the wealthy?

- Could the widespread use of gene therapy make society less accepting of people who are different?

- Should people be allowed to use gene therapy to enhance basic human traits such as height, intelligence, or athletic ability?

Current gene therapy research has focused on treating individuals by targeting the therapy to body cells such as bone marrow or blood cells. This type of gene therapy cannot be passed on to a person's children. Gene therapy could be targeted to egg and sperm cells (germ cells), however, which would allow the inserted gene to be passed on to future generations. This approach is known as germline gene therapy.

The idea of germline gene therapy is controversial. While it could spare future generations in a family from having a particular genetic disorder, it might affect the development of a fetus in unexpected ways or have long-term side effects that are not yet known. Because people who would be affected by germline gene therapy are not yet born, they can't choose whether to have the treatment. Because of these ethical concerns, the U.S. Government does not allow federal funds to be used for research on germline gene therapy in people.

For more information about the ethical issues raised by gene therapy:

The Genetics and Public Policy Center outlines scientific issues and ethical concerns regarding gene therapy (http://www.dnapolicy.org/science.gm.php).

The National Human Genome Research Institute discusses scientific issues and ethical concerns surrounding germline gene therapy (http://www.genome.gov/10004764).

A discussion of the ethics of gene therapy and genetic enhancement (http://ethics.missouri.edu/Gene-Therapy.aspx) is available from the University of Missouri Center for Health Ethics.

Is gene therapy available to treat my disorder?

Gene therapy is currently available only in a research setting. The U.S. Food and Drug Administration (FDA) has not yet approved any gene therapy products for sale in the United States.

Hundreds of research studies (clinical trials) are under way to test gene therapy as a treatment for genetic conditions, cancer, and HIV/AIDS. If you are interested in participating in a clinical trial, talk with your doctor or a genetics professional about how to participate.

You can also search for clinical trials online. ClinicalTrials.gov (http://clinicaltrials.gov/), a service of the National Institutes of Health, provides easy access to information on clinical trials. You can search for specific trials or browse by condition or trial sponsor. You may wish to refer to a list of gene therapy trials (http://clinicaltrials.gov/search?term=%22gene+therapy%22) that are accepting (or will accept) participants.

Chapter 8
The Human Genome Project

Table of Contents

What is a genome?

A genome is an organism's complete set of DNA, including all of its genes. Each genome contains all of the information needed to build and maintain that organism. In humans, a copy of the entire genome—more than 3 billion DNA base pairs—is contained in all cells that have a nucleus.

For more information about genomes:

The U.S. Department of Energy Office of Science provides background information about the human genome in its fact sheet The Science Behind the Human Genome Project (http://web.ornl.gov/sci/techresources/Human_Genome/project/info.shtml).

The Wellcome Trust Sanger Institute offers basic and in-depth explanations of the genome (http://www.yourgenome.org/dgg/general/).

What was the Human Genome Project and why has it been important?

The Human Genome Project was an international research effort to determine the sequence of the human genome and identify the genes that it contains. The Project was coordinated by the National Institutes of Health and the U.S. Department of Energy. Additional contributors included universities across the United States and international partners in the United Kingdom, France, Germany, Japan, and China. The Human Genome Project formally began in 1990 and was completed in 2003, 2 years ahead of its original schedule.

The work of the Human Genome Project has allowed researchers to begin to understand the blueprint for building a person. As researchers learn more about the functions of genes and proteins, this knowledge will have a major impact in the fields of medicine, biotechnology, and the life sciences.

For more information about the Human Genome Project:

The National Human Genome Research Institute offers a fact sheet about the Human Genome Project (http://www.genome.gov/10001772) and a list of frequently asked questions (http://www.genome.gov/11006943). Additionally, the booklet From the Blueprint to You provides an overview of the project (http://www.genome.gov/Pages/Education/Modules/BluePrintToYou/Blueprint7to8.pdf).

A brief description of the Project and links to many additional resources are available from the Human Genome Project Information web site (http://www.ornl.gov/sci/techresources/Human_Genome/home.shtml), a service of the U.S. Department of Energy Office of Science.

An overview of the Human Genome Project (http://www.genetics.edu.au/Publications-and-Resources/Genetics-Fact-Sheets/FactSheet24) is available from the Centre for Genetics Education.

Additional information can be found in the National Library of Medicine fact sheet Understanding the Human Genome Project (http://www.nlm.nih.gov/medlineplus/magazine/issues/summer13/articles/summer13pg15.html).

The Wellcome Trust Sanger Institute offers and overview of the Human Genome Project, including its history (http://www.yourgenome.org/downloads/pdf/hgp/hgp.pdf) and approaches to sequencing (http://www.yourgenome.org/hgp/).

The Smithsonian National Museum of Natural History's exhibit 'Genome: Unlocking Life's Code' provides a timeline (http://unlockinglifescode.org/timeline) of the Human Genome Project.

What were the goals of the Human Genome Project?

The main goals of the Human Genome Project were to provide a complete and accurate sequence of the 3 billion DNA base pairs that make up the human genome and to find all of the estimated 20,000 to 25,000 human genes. The Project also aimed to sequence the genomes of several other organisms that are important to medical research, such as the mouse and the fruit fly.

In addition to sequencing DNA, the Human Genome Project sought to develop new tools to obtain and analyze the data and to make this information widely available. Also, because advances in genetics have consequences for individuals and society, the Human Genome Project committed to exploring the consequences of genomic research through its Ethical, Legal, and Social Implications (ELSI) program.

For more information about the Human Genome Project's goals:

The National Human Genome Research Institute provides a fact sheet about DNA sequencing (http://www.genome.gov/10001177).

The National Human Genome Research Institute details the goals and accomplishments (https://www.genome.gov/11006945) of the Human Genome Project.

What did the Human Genome Project accomplish?

In April 2003, researchers announced that the Human Genome Project had completed a high-quality sequence of essentially the entire human genome. This sequence closed the gaps from a working draft of the genome, which was published in 2001. It also identified the locations of many human genes and provided information about their structure and organization. The Project made the sequence of the human genome and tools to analyze the data freely available via the Internet.

In addition to the human genome, the Human Genome Project sequenced the genomes of several other organisms, including brewers' yeast, the roundworm, and the fruit fly. In 2002, researchers announced that they had also completed a working draft of the mouse genome. By studying the similarities and differences between human genes and those of other organisms, researchers can discover the functions of particular genes and identify which genes are critical for life.

The Project's Ethical, Legal, and Social Implications (ELSI) program became the world's largest bioethics program and a model for other ELSI programs worldwide. For additional information about ELSI and the program's accomplishments, please refer to What were some of the ethical, legal, and social implications addressed by the Human Genome Project? (http://ghr.nlm.nih.gov/handbook/hgp/elsi)

For more information about the accomplishments of the Human Genome Project:

An overview of the Project's accomplishments is available in the National Human Genome Research Institute news release International Consortium Completes Human Genome Project (http://www.genome.gov/11006929).

A 2004 news release (http://www.genome.gov/12513430) about the finished human genome sequence is available from the National Human Genome Research Institute.

What were some of the ethical, legal, and social implications addressed by the Human Genome Project?

The Ethical, Legal, and Social Implications (ELSI) program was founded in 1990 as an integral part of the Human Genome Project. The mission of the ELSI program was to identify and address issues raised by genomic research that would affect individuals, families, and society. A percentage of the Human Genome Project budget at the National Institutes of Health and the U.S. Department of Energy was devoted to ELSI research.

The ELSI program focused on the possible consequences of genomic research in four main areas:

- Privacy and fairness in the use of genetic information, including the potential for genetic discrimination in employment and insurance.

- The integration of new genetic technologies, such as genetic testing, into the practice of clinical medicine.

- Ethical issues surrounding the design and conduct of genetic research with people, including the process of informed consent (http://ghr.nlm.nih.gov/handbook/testing/informedconsent).

- The education of healthcare professionals, policy makers, students, and the public about genetics and the complex issues that result from genomic research.

For more information about the ELSI program:

Information about the ELSI program at the National Institutes of Health, including program goals and activities, is available in the fact sheet Ethical, Legal and Social Implications (ELSI) Research Program (http://www.genome.gov/10001618) from the National Human Genome Research Institute. The ELSI Planning and Evaluation History web page (http://www.genome.gov/10001754) provides a more detailed discussion of the program.

More discussion about ethical issues in human genetics (http://www.genetics.edu.au/Publications-and-Resources/Genetics-Fact-Sheets/FactSheet23), including genetic discrimination, the cloning of organisms, and the patenting of genes is available from the Centre for Genetics Education.

The World Health Organization provides further discussion of the ELSI implications (http://www.who.int/genomics/elsi/en/) of human genomic research.

Chapter 9
Genomic Research

Table of Contents

What are the next steps in genomic research?

Discovering the sequence of the human genome was only the first step in understanding how the instructions coded in DNA lead to a functioning human being. The next stage of genomic research will begin to derive meaningful knowledge from the DNA sequence. Research studies that build on the work of the Human Genome Project are under way worldwide.

The objectives of continued genomic research include the following:

- Determine the function of genes and the elements that regulate genes throughout the genome.

- Find variations in the DNA sequence among people and determine their significance. The most common type of genetic variation is known as a single nucleotide polymorphism or SNP (pronounced "snip"). These small differences may help predict a person's risk of particular diseases and response to certain medications.

- Discover the 3-dimensional structures of proteins and identify their functions.

- Explore how DNA and proteins interact with one another and with the environment to create complex living systems.

- Develop and apply genome-based strategies for the early detection, diagnosis, and treatment of disease.

- Sequence the genomes of other organisms, such as the rat, cow, and chimpanzee, in order to compare similar genes between species.

- Develop new technologies to study genes and DNA on a large scale and store genomic data efficiently.

- Continue to explore the ethical, legal, and social issues raised by genomic research.

For more information about the genomic research following the Human Genome Project:

The National Human Genome Research Institute supports research in many of the areas described above. The Institute provides detailed information about its research initiatives at NIH (http://www.genome.gov/ResearchAtNHGRI/).

The U.S. Department of Energy Office of Science provides information about its genomics programs at genomics.energy.gov (http://genomics.energy.gov/) and a timeline of research events (http://www.ornl.gov/sci/techresources/Human_Genome/project/timeline.shtml) during and since the Human Genome Project.

The Genome Institute at Washington University explains the 1000 Genomes Project (http://genome.wustl.edu/projects/detail/1000-genomes-project/), which furthers the work of the HapMap Project.

The Wellcome Trust Sanger Institute discusses the 1000 Genomes Project (http://www.wellcome.ac.uk/Funding/Biomedical-science/Funded-projects/Major-initiatives/WTDV029748.htm) in a video that describes the key findings of the project.

What are single nucleotide polymorphisms (SNPs)?

Single nucleotide polymorphisms, frequently called SNPs (pronounced "snips"), are the most common type of genetic variation among people. Each SNP represents a difference in a single DNA building block, called a nucleotide. For example, a SNP may replace the nucleotide cytosine (C) with the nucleotide thymine (T) in a certain stretch of DNA.

SNPs occur normally throughout a person's DNA. They occur once in every 300 nucleotides on average, which means there are roughly 10 million SNPs in the human genome. Most commonly, these variations are found in the DNA between genes. They can act as biological markers, helping scientists locate genes that are associated with disease. When SNPs occur within a gene or in a regulatory region near a gene, they may play a more direct role in disease by affecting the gene's function.

Most SNPs have no effect on health or development. Some of these genetic differences, however, have proven to be very important in the study of human health. Researchers have found SNPs that may help predict an individual's response to certain drugs, susceptibility to environmental factors such as toxins, and risk of developing particular diseases. SNPs can also be used to track the inheritance of disease genes within families. Future studies will work to identify SNPs associated with complex diseases such as heart disease, diabetes, and cancer.

For more information about SNPs:

An audio definition of SNPs (http://www.genome.gov/glossary/?id=185) is available from the National Human Genome Research Institute's Talking Glossary of Genetic Terms.

How scientists locate SNPs in the genome (http://learn.genetics.utah.edu/content/pharma/snips/) is explained by the University of Utah Genetics Science Learning Center.

For people interested in more technical data, several databases of known SNPs are available:

- NCBI database of single nucleotide polymorphisms (dbSNP) (http://www.ncbi.nlm.nih.gov/SNP/)

- Database of Japanese single nucleotide polymorphisms (JSNP) (http://snp.ims.u-tokyo.ac.jp/)

What are genome-wide association studies?

Genome-wide association studies are a relatively new way for scientists to identify genes involved in human disease. This method searches the genome for small variations, called single nucleotide polymorphisms or SNPs (pronounced "snips"), that occur more frequently in people with a particular disease than in people without the disease. Each study can look at hundreds or thousands of SNPs at the same time. Researchers use data from this type of study to pinpoint genes that may contribute to a person's risk of developing a certain disease.

Because genome-wide association studies examine SNPs across the genome, they represent a promising way to study complex, common diseases in which many genetic variations contribute to a person's risk. This approach has already identified SNPs related to several complex conditions including diabetes, heart abnormalities, Parkinson disease, and Crohn disease. Researchers hope that future genome-wide association studies will identify more SNPs associated with chronic diseases, as well as variations that affect a person's response to certain drugs and influence interactions between a person's genes and the environment.

For more information about genome-wide association studies:

The National Human Genome Research Institute provides a detailed explanation of genome-wide association studies (http://www.genome.gov/20019523).

Nature Education's Scitable provides an explanation of genome-wide association studies and how they assist in estimating the risk (http://www.nature.com/scitable/topicpage/genetic-variation-and-disease-gwas-682) of disease.

Genomics Unzipped explains how to read a genome-wide association study (http://www.genomesunzipped.org/2010/07/how-to-read-a-genome-wide-association-study.php).

You can also search for clinical trials of genome-wide association studies online. ClinicalTrials.gov (http://clinicaltrials.gov/), a service of the National Institutes of Health, provides easy access to information on clinical trials. You can search for specific trials or browse by condition or trial sponsor. You may wish to refer to a list of genome-wide association studies (http://clinicaltrials.gov/search?term=GWAS+OR+%22Genome+Wide+Association%22) that are accepting (or will accept) participants.

For people interested in more technical information, the NCBI's Database of Genotype and Phenotype (dbGaP) (http://www.ncbi.nlm.nih.gov/sites/entrez?db=gap) contains data from genome-wide association studies. An introduction to this database, as well as information about study results, is available from the dbGaP press release (http://www.nlm.nih.gov/archive//20120510/news/press_releases/

dbgap_launchPR06.html). In addition, the National Human Genome Research Institute provides a Catalog of Published Genome-Wide Association Studies (http://www.genome.gov/gwastudies/).

What is the International HapMap Project?

The International HapMap Project is an international scientific effort to identify common genetic variations among people. This project represents a collaboration of scientists from public and private organizations in six countries. Data from the project is freely available to researchers worldwide. Researchers can use the data to learn more about the relationship between genetic differences and human disease.

The HapMap (short for "haplotype map") is a catalog of common genetic variants called single nucleotide polymorphisms or SNPs (pronounced "snips"). Each SNP represents a difference in a single DNA building block, called a nucleotide. These variations occur normally throughout a person's DNA. When several SNPs cluster together on a chromosome, they are inherited as a block known as a haplotype. The HapMap describes haplotypes, including their locations in the genome and how common they are in different populations throughout the world.

The human genome contains roughly 10 million SNPs. It would be difficult, time-consuming, and expensive to look at each of these changes and determine whether it plays a role in human disease. Using haplotypes, researchers can sample a selection of these variants instead of studying each one. The HapMap will make carrying out large-scale studies of SNPs and human disease (called genome-wide association studies) cheaper, faster, and less complicated.

The main goal of the International HapMap Project is to describe common patterns of human genetic variation that are involved in human health and disease. Additionally, data from the project will help researchers find genetic differences that can help predict an individual's response to particular medicines or environmental factors (such as toxins.)

For more information about the International HapMap Project:

The National Human Genome Research Institute provides an overview of the project in their International HapMap Project fact sheet (http://www.genome.gov/10001688). The fact sheet also includes a link to a more in-depth online tutorial on HapMap usage.

Detailed information about the project, as well as project data, are available from the International HapMap Project web site (http://hapmap.ncbi.nlm.nih.gov/).

Cold Springs Harbor Laboratory provides a step-by-step guide (http://cshlpress.com/pdf/sample/GenVa-06.pdf) to using the International HapMap Project web site.

The Ohio Channel, a service of Ohio's public broadcasting stations, posted a video in which an overview (http://www.ohiochannel.org/MediaLibrary/Media.aspx?fileId=122803) of the International HapMap Project is described.

You can also search for clinical trials involving haplotypes or associated with the International HapMap Project. ClinicalTrials.gov (http://clinicaltrials.gov/), a service of the National Institutes of Health, provides easy access to information on clinical trials. You can search for specific trials or browse by condition or trial sponsor. You may wish to refer to a list of haplotype-related studies (http://clinicaltrials.gov/search?term=HAPMAP+OR+haplotype) that are accepting (or will accept) participants.

What is the Encyclopedia of DNA Elements (ENCODE) Project?

The ENCODE Project was planned as a follow-up to the Human Genome Project. The Human Genome Project sequenced the DNA that makes up the human genome; the ENCODE Project seeks to interpret this sequence. Coinciding with the completion of the Human Genome Project in 2003, the ENCODE Project began as a worldwide effort involving more than 30 research groups and more than 400 scientists.

The approximately 20,000 genes that provide instructions for making proteins account for only about 1 percent of the human genome. Researchers embarked on the ENCODE Project to figure out the purpose of the remaining 99 percent of the genome. Scientists discovered that more than 80 percent of this non-gene component of the genome, which was once considered "junk DNA," actually has a role in regulating the activity of particular genes (gene expression).

Researchers think that changes in the regulation of gene activity may disrupt protein production and cell processes and result in disease. A goal of the ENCODE Project is to link variations in the expression of certain genes to the development of disease.

The ENCODE Project has given researchers insight into how the human genome functions. As researchers learn more about the regulation of gene activity and how genes are expressed, the scientific community will be able to better understand how the entire genome can affect human health.

For more information about the ENCODE Project:

The University of California at Santa Cruz provides detailed information about the findings of the ENCODE Project (http://encodeproject.org/ENCODE/) as well as the Project's experimental procedures and many other types of data.

Published research findings are available through Nature Magazine's Nature Encode Explorer (http://www.nature.com/encode/#/threads), which gives the public access to scientific information collected from the ENCODE Project.

The Broad Institute describes the purpose (http://www.broadinstitute.org/news/4326) of the ENCODE Project.

The National Human Genome Research Institute announces results of the ENCODE Project in a Press Release (http://www.genome.gov/27549810) and provides an overview (http://www.genome.gov/10005107) of the ENCODE Project.

Some of the outcomes of the ENCODE Project (http://www.cshl.edu/news-a-features/massive-genome-analysis-by-encode-redefines-the-gene-and-sheds-new-light-on-complex-disease.html) are detailed by Cold Springs Harbor Laboratory.

What is pharmacogenomics?

Pharmacogenomics is the study of how genes affect a person's response to drugs. This relatively new field combines pharmacology (the science of drugs) and genomics (the study of genes and their functions) to develop effective, safe medications and doses that will be tailored to a person's genetic makeup.

Many drugs that are currently available are "one size fits all," but they don't work the same way for everyone. It can be difficult to predict who will benefit from a medication, who will not respond at all, and who will experience negative side effects (called adverse drug reactions). Adverse drug reactions are a significant cause of hospitalizations and deaths in the United States. With the knowledge gained from the Human Genome Project, researchers are learning how inherited differences in genes affect the body's response to medications. These genetic differences will be used to predict whether a medication will be effective for a particular person and to help prevent adverse drug reactions.

The field of pharmacogenomics is still in its infancy. Its use is currently quite limited, but new approaches are under study in clinical trials. In the future, pharmacogenomics will allow the development of tailored drugs to treat a wide range of health problems, including cardiovascular disease, Alzheimer disease, cancer, HIV/AIDS, and asthma.

For more information about pharmacogenomics

The National Institute of General Medical Sciences offers a list of Frequently Asked Questions about Pharmacogenomics (http://www.nigms.nih.gov/Research/SpecificAreas/PGRN/Background/pages/pgrn_faq.aspx).

A list of Frequently Asked Questions about Pharmacogenomics (https://www.genome.gov/27530645) is also offered by the National Human Genome Research Institute.

Additional information about pharmacogenetics is available from the Centre for Genetics Education (http://www.genetics.edu.au/Publications-and-Resources/Genetics-Fact-Sheets/FactSheet25) as well as Genes In Life (http://www.genesinlife.org/testing-services/testing-genetic-conditions/pharmacogenomic-testing).

The Smithsonian National Museum of Natural History's exhibit 'Genome: Unlocking Life's Code' discusses the utility of pharmacogenomics (http://unlockinglifescode.org/explore/genomic-medicine/pharmacogenomics).

The Genetic Science Learning Center at the University of Utah offers an interactive introduction to pharmacogenomics (http://learn.genetics.utah.edu/content/pharma/).

Another interactive tutorial (http://www.phgfoundation.org/tutorials/pharmacogenomics/) is available from the PHG Foundation.

The American Medical Association explains what pharmacogenomics is and provides a list of practical applications (http://www.ama-assn.org/ama/pub/physician-resources/medical-science/genetics-molecular-medicine/current-topics/pharmacogenomics.page).

The National Genetics and Genomics Education Centre of the National Health Service (UK) provides information about predicting the effects of drugs (http://www.geneticseducation.nhs.uk/genomics-in-health/predict-drug-effects) based on a person's genes.

PharmGKB (https://www.pharmgkb.org/) is a pharmacogenomics resource sponsored by the National Institutes of Health that collects information on human genetic variation and drug responses.

A list of clinical trials involving pharmacogenomics (http://clinicaltrials.gov/search/term=pharmacogenomics+OR+pharmacogenetics) is available from ClinicalTrials.gov, a service of the National Institutes of Health.

What advances are being made in DNA sequencing?

Determining the order of DNA building blocks (nucleotides) in an individual's genetic code, called DNA sequencing, has advanced the study of genetics and is one method used to test for genetic disorders.

New technologies that allow rapid sequencing of large amounts of DNA are being developed. The original sequencing technology, called Sanger sequencing (named after the scientist who developed it, Frederick Sanger), was a breakthrough that helped scientists determine the human genetic code, but it is time-consuming and expensive. The Sanger method has been automated to make it faster and is still used in laboratories today to sequence short pieces of DNA, but it would take years to sequence all of a person's DNA (known as the person's genome). Several technologies have been developed more recently, called next-generation sequencing (or next-gen sequencing), that have sped up the process (taking only days to weeks to sequence a human genome) while reducing the cost.

With next-generation sequencing, it is now feasible to sequence large amounts of DNA, for instance all the pieces of an individual's DNA that provide instructions for making proteins. These pieces, called exons, are thought to make up 1 percent of a person's genome. Together, all the exons in a genome are known as the exome, and the method of sequencing them is known as whole exome sequencing. This method allows variations in the protein-coding region of any gene to be identified, rather than a select few genes. Because most known mutations that cause disease occur in exons, whole exome sequencing is thought to be an efficient method to identify possible disease-causing mutations.

However, researchers have found that DNA variations outside the exons can affect gene activity and protein production and lead to genetic disorders–variations that whole exome sequencing would miss. Another method, called whole genome sequencing, determines the order of all the nucleotides in an individual's DNA and can determine variations in any part of the genome.

While many more genetic changes can be identified with whole exome and whole genome sequencing than with select gene sequencing, the significance of much of this information is unknown. Because not all genetic changes affect health, it is difficult to know whether identified variants are involved in the condition of interest. Sometimes, an identified variant is associated with a different genetic disorder that has not yet been diagnosed (these are called incidental or secondary findings).

In addition to being used in the clinic, whole exome and whole genome sequencing are valuable methods for researchers. Continued study of exome and genome sequences can help determine whether new genetic variations are associated with health conditions, which will aid disease diagnosis in the future.

For more information about DNA sequencing technologies and their use:

Genetics Home Reference discusses whether all genetic changes affect health and development (http://ghr.nlm.nih.gov/handbook/mutationsanddisorders/ neutralmutations).

A scientist at the Genome Institute at the University of Washington describes the different sequencing technologies (http://genome.wustl.edu/articles/detail/dna-sequencing-technology-a-perspective-from-dr-elaine-mardis/) and what the new technologies have meant for the study of the genetic code.

An illustration of the decline in the cost of DNA sequencing (http://www.genome.gov/ sequencingcosts/), including that caused by the introduction of new technologies, is provided by the National Human Genome Research Institute (NHGRI).

The American College of Medical Genetics and Genomics (ACMG) has laid out their policies regarding whole exome and whole genome sequencing (http://www.acmg.net/StaticContent/PPG/Clinical_Application_of_Genomic_ Sequencing.pdf), including when these methods should be used, what results may arise, and what the results might indicate.

The PHG Foundation (UK) provides an overview of whole genome sequencing (http://www.phgfoundation.org/file/10365/) and how it can be used in healthcare.

The Mount Sinai School of Medicine Genomics Core Facility describes the techniques used in whole exome sequencing (http://www.mssm.edu/research/ institutes/genomics-institute/genomics-core-facility/genomic-technology-applications/ whole-exome-sequencing).

The Smithsonian National Museum of Natural History's exhibit 'Genome: Unlocking Life's Code' discusses the advancements (http://unlockinglifescode.org/explore/ genome-within-us/reading-lifes-code) made in DNA sequencing.